D1321677

THE NEW NORMAL

THE NEW NORMAL

PETER HINSSEN

THE
NEW
NORMAL

EXPLORE THE LIMITS
OF THE DIGITAL WORLD

With contributions by Misha Chellam

ACROSS
TECHNOLOGY

THE NEW NORMAL

EXPLORE THE LIMITS OF THE DIGITAL WORLD

Publisher n°: 73320 Mach Media NV

Technologiepark 3, B- 9052 Gent, Belgium

Tel. +32 9 243 60 11 – www.machmedia.be

ISBN 9789081324250 – 4rd edition

Legal Deposit: D/2010/11.651/5

Printed in Belgium

AUTHOR	Peter Hinssen
CONTRIBUTOR	Misha Chellam
PROJECT MANAGERS	Luc Osselaer and Marianne Vermeulen
CREATION	Steven Theunis, www.armeedeverre.be
ILLUSTRATIONS	Vera Ponnet, Saflot

I would like to thank all the people who were absolutely vital in making this book a reality
(in alphabetical order): Courtney Davis, Ilse De Bondt, José Delameilleure, Carole Gheysen,
Delphine Hajaji, Luc Osselaer, Taunya Renson-Martin, Maarten Van Steenbergen,
Marianne Vermeulen, Sybylla Wales.

I would also like to thank all the CEOs, CIOs, business partners and IT thought leaders that were a
source of inspiration and motivation through hundreds of wonderful conversations (in alphabetical
order): Patrick Arlequeeuw, Jean-Michel Binard, Abdella Bouharrak, Alex Brabers, Neil Cameron,
Francis Colardyn, LaVerne Council, Susan Cramm, Eric Cuypers, Paul Danneels, Nicholas Davies,
Christiaan De Backer, Herman De Prins, Bart De Ruijter, Kurt De Ruwe, Geert Desmet, Alain De Taeye,
Erik Dralans, Koen Fruyt, Robert Goffee, Ajei Gopal, Philippe Gosseye, Christophe Jacques,
Michael Kögeler, Geert Mareels, Costas Markides, Eddy Minnaert, Philippe Naert, Philip Neyt,
Geert Noels, Carlota Perez, Kosta Peric, Patrick Reyniers, Menno Rientjes, Matteo Rizzi,
Fonny Schenck, Tina Seelig, Hans Tesselaar, Saul van Beurden, André Vanden Camp,
Bart van den Bosch, Ludo Van den Kerckhove, Peter Vander Auwera, Olivier van der Brempt,
Ton van der Linden, Ludo Van der Velden, Peter Vandevenne, Geert Van Hove, Bart Vannieuwenhuyse,
Vincent Van Quickenborne, Patrick Van Renterghem, Geert Van Wonterghem, Steve Van Wyk,
Luc Verhelst, Ludo Wijckmans.

Copyright pictures:
With special thanks to Philips for the visionary image of ambient intelligence: p.65;
Corbis: p. 97, 151; iStock: p. 1, 11, 16, 22, 23, 25, 27, 33, 39, 45, 47, 50, 55, 58, 59, 61, 63, 95, 96, 116, 123, 127,
130, 145, 147, 155, 164, 167, 172, 173, 189, 197, 208.

www.neonormal.com   www.peterhinssen.com
www.lannoo.com   www.a-cross.com

TO MY PARENTS

To my mother,
who gifted me with a love of
language, a joy for words,
and a passion for writing.

To my father,
who instilled in me a love of
technology and who bought me
my first computer when I was
only 13... Although I still wish you
had bought the Apple II like I asked.

Peter

CONTENTS

9   Preface

10   CHAPTER **01** INTRODUCING THE NEW NORMAL

12   Halfway there

15   A brief history of digital

16   Digital immigrants vs. digital natives

17   The New Normal in the workplace

19   The four I's

23   Can you adapt?

26   CHAPTER **02** EXPLORING THE LIMITS

28   What's a limit anyway?

29   The limit of length

33   The limit of depth

35   The limit of price

37   The limit of patience

39   The limit of privacy

41   The limit of intelligence

43   So, what are your limits?

44   CHAPTER **03** RULES OF THE NEW NORMAL

46   New rules

46   Rule #1: Zero tolerance for digital failure

49   Rule #2: Good enough beats perfect

53   Rule #3: The era of total accountability

57   Rule #4: Abandon absolute control

62   Conclusion

64   CHAPTER **04** CUSTOMER STRATEGIES FOR THE NEW NORMAL

66   From info-bahn to You

69   What we can learn from the dot-com bubble

78   Contact is king. And the user is the dictator.

81   The experience economy

82   The new patterns

93   Conclusion

94   CHAPTER **05**   INFORMATION STRATEGIES FOR THE NEW NORMAL

96   The value of information

100   From puzzles to mysteries

103   The basics

109   Quantum thinking in information

110   Drivers of information strategy

116   It's not information overload; it's filter failure

121   Conclusion

122   CHAPTER **06**   DESCRIBING ORGANIZATIONS IN THE NEW NORMAL

124   Total accountability zooms in

128   Intrapreneurship

134   The boundaries of the New Normal

137   Entrepreneurs in the New Normal

139   The company and you

143   Conclusion

144   CHAPTER **07**   INNOVATION IN THE NEW NORMAL

146   The seeds of innovation

147   And then came digital

151   The PARC dilemma

155   Playing chess vs. playing poker

156   From R&D to C&D, Connect & Develop

158   Open innovation

164   Conclusion

166   CHAPTER **08**   TECHNOLOGY STRATEGY FOR THE NEW NORMAL

168   Technology strategy for the New Normal

169   From build, to buy, to compose

176   Commoditization and the cloud

179   Multi-tenancy and rethinking IT

186   Conclusion

188   CHAPTER **09**   THE BIG PICTURE

190   Recap

195   We're halfway there

195   So what happens next?

197   Through the eyes of our children

201   Sources

HE AND SHE

Throughout this book,
I will talk about consumers, CIOs,
users... referring to them as 'he'.
Does this mean that this book
is 'men only'? On the contrary!
I sincerely hope there will be as
many women as men thriving
in the New Normal.
So my only bias is that I think
using 'he or she' is tedious to read.

PREFACE

Do you remember when you heard the word 'digital' for the first time?

If you're a nerd, it was probably quite a way back, and it was probably being talked about in ones and zeroes.

For the rest of you, was it your father talking about a 'digital watch'? Remember those Seiko timepieces from the 70s?

Was it perhaps when you held your first Compact Disc, after carefully lifting it from its CD box like it was made of titanium, while friends commented that the sound was soooo much better than a vinyl record because it was 'digital'?

Or was it when you first took a picture with a 'digital camera' and were amazed to be able to immediately view the image you had taken?

Or was it when you learned about 'digital television', or 'digital radio', or...

There was surely a moment when the word 'digital' entered your consciousness for the very first time.

Today, we are entering an era where 'digital' has become an almost useless adjective.

Analog is what's really special nowadays.

Analog watches are vintage items, and the absolute sign of wealth is to have an old antique Breguet on your wrist to demonstrate status and sophistication.

True diehards are scavenging used record stores in search of collector edition vinyl records and LPs, and are playing them through old 'transitor tubes' analog amplifiers because they want to recreate that true analog sound.

Analog is cool because the world has gone truly digital. Your kids don't even think about the fact that things are digital, they just assume that everything is.

BUT. It would be false to assume that now that we've 'arrived' in the digital world, that's the end of that. As a matter of fact, we're just about to embark on an exciting second stage of the journey into the digital revolution.

Peter Hinssen
July 2010

INTRODUCING
THE NEW NORMAL

"If a man will begin with certainties,
   he shall end in doubts.
   But if he will be content to begin with doubts,
   he shall end in certainties."
— Francis Bacon

"History doesn't repeat itself.
   But it does rhyme."
— Mark Twain

We're at the halfway mark in the digital revolution.
What will happen when we cross this point?

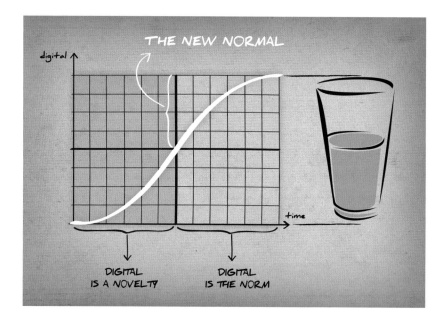

The idea behind **the New Normal** is quite simple: 'We're halfway there'. The New Normal is about all things we call 'digital', and in the digital revolution we're probably only halfway there. That means we have as much journey ahead of us as we have behind us.

In this book, we want to explore the world of 'digital' once we've crossed the halfway point, specifically examining the impact of the New Normal on our business lives. In our opinion, the second leg of the digital journey will be completely different from the first leg that is now coming to a close.

The glass is half full, and this will drastically change our perception of technology. The next ten, twenty and thirty years will be marked by an ever increasing digital world and society, but there will be some fundamental shifts in our behavior and our adoption of technology. The impact on businesses and the way they use technology will be enormous.

How do we know that we're halfway there? We don't. Justin Rattner is Vice President and Chief Technology Officer of Intel, the company that powers most of our computers with their Intel

Justin R. Rattner
www.intel.com

chips. Justin is a 'deep' thinker, and his point of view is that we've now had roughly 40 years of Information Technology, but as Justin says: "If you think the past years of digital revolution were pretty amazing, think again. The next 40 years will blow you away, and will make the past 40 years look pretty tame."

Quite a bold statement, but then again, Intel builds and sells the technology of the future. The future is being made in their labs at this very moment, so he should know.

One of the co-founders of Intel, Gordon Moore, famously identified a trend in computing power that has been shap-

Gordon E. Moore
www.intel.com

ing the industry since the 1960s. According to 'Moore's Law', the power and speed of computers double approximately every two years, and the semi-conductor industry now sets targets for research and development based on exponential gains in digital capac-ity about every two years.

Moore's Law: Capacity in IT doubles about every 2 years.

Ray Kurzweil, the great future thinker, put this relentless trend in context: "The computation in a cell phone today is a

Ray Kurzweil
www.kurzweiltech.com

million times cheaper and a thousand times more powerful than the com-puter I used as a student. That's a billion-fold increase in price performance. And it ain't about to stop now."

Kurzweil is notoriously optimistic about the upcoming decades of the digital revolution, but this is not a book about how great and wonderful the digital future is going to be. We have focused here on the impact of the dig-ital revolution on your business, and how the awareness of the trends that we've seen over the past generation can help shape your business strategy going forward.

THE NEW NORMAL

DIGITAL IS THE NEW BLACK.

What will happen is that digital will become the New Normal. For more than thirty years, we've become increasingly digital, and the moment has arrived that we're passing the halfway mark. From now on, digital is the norm. Everything we do from this point forward will have one common characteristic; we will *expect* things to be digital.

This is a statement on how our perception changes when digital is just 'normal'. How we will expect digital things to always work, and will have a zero tolerance for digital failure.

This is about how we will crave for human contact in a world where most interactions with organizations and companies will be digital, and where analog might actually see a revival.

We will have to market to consumers differently when the world has become a digital society with the user in control of the consumption of information.

The second leg of the digital revolution won't be a walk in the park. In fact, it will be much more difficult to truly stand out and differentiate your organization. In the beginning, it was simple; using technology was a novelty and gave you an edge. In the New Normal, access to technology is a commodity, and you will have to focus on other skills in your organization to make the difference.

The analogy I like to use is the swimming pool. If the digital revolution is the complete pool, it would be safe to say that you're halfway in the water. You've waded through the shallow end, but your feet still touch the ground of the pool. There is no indication, however, that you will be able to swim in the deep end.

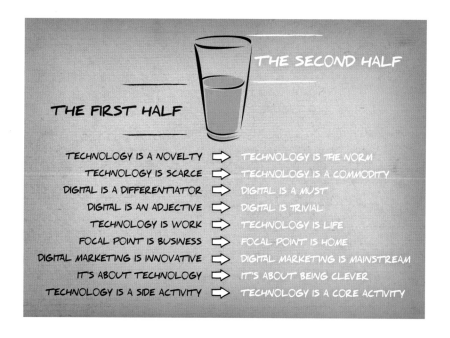

IT departments will have to adapt to a world where they no longer introduce hot new technology, but where they will have to play catch-up to what their employees are using in their homes and find completely natural.

Let's now explore the deep end of the digital pool.

A BRIEF HISTORY OF DIGITAL

The roots of the digital revolution stretch back over 2000 years, to the invention of the abacus and the slide rule, and wind throughout the following two millennia, tracing innovations in both automated calculation and programmability. Fast forward to the 20th century, when, over a ten-year period during the 1930s and 40s, the cumulative innovations of a wide range of scientists and engineers led to the birth of modern digital computing, driven in large part by government-funded research.

The 1960s saw the transition from vacuum-tube-based electronics to transistor-based machines, which ushered in a new paradigm of digital advancement, framed broadly by 'Moore's Law'. As speed increased and the need for power decreased, the digital world spread from governments to corporations to individuals. And at a certain point at the end of the 1990s, the decreased costs of hardware combined with the growing World Wide Web and a consumer-centric software revolution caused an inflection point:

The technology in our home eclipsed the technology in our work life.

SOCIETY HAS CLEARLY GONE DIGITAL
IT AIN'T KANSAS ANYMORE. IT'S SILICON.

Observe the changes in our home environment.

Ten years ago, if you had access to a laptop, it was probably on loan from your IT department. And ten years ago, this laptop, or your desktop PC, was probably the nicest, most expensive, most sophisticated piece of technology in your home, and was gradually handed down from the older generation to the newer generation. In previous times, clothes were handed down, but technology became the new hand-me-down.

Today, the tables have turned: the most powerful technology in the house may be owned by a teenager who needs the 'fastest and coolest' computer in the house to play 'World of Warcraft'. When that won't do anymore, it will be handed over to the siblings, and eventually you end up with something on which you do your home banking. That is a total reversal of power in our living rooms in just ten years.

I saw a recent survey in Australia on the necessities in life. They asked hundreds of (young) Australians what they thought were the most important 'necessities'. The top two answers – ① a car, and ② air conditioning – were not surprising, as Australia is a big, hot continent. But the rest of the top 10 list was dominated by digital fare like iPods, laptops, Facebook, high speed Internet, mobile phones and flat screen TVs. As a sign of the times, on the top ten list of the absolute 'necessities in life', food didn't make the cut.

Physiologically speaking, these young Australians are jumping the gun a bit, but we can clearly see the perceptions of digital importance in the younger generation.

DIGITAL IMMIGRANTS VS. DIGITAL NATIVES

It's a camera of course!

Most readers of this book don't have a long history of being digital. Most of us are still digital immigrants. But we're entering a world where analog really is the exception. I like this simple test to distinguish between a digital native and a digital immigrant. You put a camera on a table and just ask "What is this?" A digital immigrant will say: "That's a *digital* camera", whereas a digital native will say: "It's a camera". A digital native has probably never used an analog camera in his life.

The more we use the adjective 'digital', the more antiquated we sound. When digital becomes the New Normal, the adjective digital loses its descriptive meaning. Perhaps we should stop using the word altogether, because it is a clear indication that we're still deeply anchored in the first leg of the journey.

THE NEW NORMAL IN THE WORKPLACE

What happens in the workplace as younger workers continue to replace older workers and as employees of all ages live increasingly digital personal lives? This is a key issue that all businesses should reflect on to make the most of opportunities within this transitional period. Today technology is a commodity. Technology is not work. Technology is life. Just a decade ago, our employees got their laptops, desktops, cell phones and Internet access from work. Today, they have better computers and connectivity at home, and they have much cooler personal cell phones than the ones we give them.

This is causing trouble for the IT folks. IT departments have turned from being 'dispensers of cool stuff' into the 'dispensers of really old boring stuff' in just a couple of years. When we used to hand out a computer and cell phone to a new employee on the first day of work, the employee would say, "Thank you". Now when we hand out their laptop and phone they look at us incredulously and think, "You don't think I'm going to be seen on the street carrying *that* around, do you?"

Generation Y – people born after 1978, typically the newest and youngest crop of office workers – are particularly affected. This is the first generation who have lived most of their lives in a digital world, and they think differently about technology. For them, technology is not for work but for life. They wake up with technology and go to sleep with technology, and as a matter of fact, their definition of *work* is: "That brief period during the day when I still have to use *old* technology."

WORK = The brief period during the day when I use old technology.

In a business context, what role will IT play in the realm of the New Normal? In my previous book – 'Business / IT Fusion' – I talked at length about the need for IT departments to transform themselves and become much closer to the business.

Peter Hinssen, *Business / IT Fusion. How to move beyond Alignment and transform IT in your organization*, 2008

But here, we're going even further. This is not about just re-positioning IT, or re-thinking the role of IT. Here we will examine how to reshape IT into something completely different. We will try to define what the 'shape under the drape' will be, and define what will happen to IT when IT becomes a commodity. We will redefine IT in the New Normal.

We are already feeling the impact of commoditization in IT. Commoditization pressures are high, and they are manifested not only in employees who are agitating to buy their own IT kits and bring them to work, but in tough questions by senior executives. Questions that all start with:

"Explain to me why"
such as:

→ "Explain to me why the upgrade of our corporate website is going to cost half a million, while my 13-year-old nephew built the website for his school last weekend on open source…"

→ "Explain to me why I can't find anything in our state-of-the-art document management system, and particularly why I had to spend over an hour retrieving a document I had put there myself, when I can go onto Google and find *anything* in less than three seconds …"

→ "Explain to me why I can book a flight online in two minutes, but it takes me almost an hour to input my expense account statements in these horrible SAP screens…"

→ "Explain to me why I can buy a printer in the shop around the corner for under $200 and have it here in fifteen minutes, but when I use the corporate IT channels it takes me two weeks and costs three times as much…"

And that is just the beginning. The power of IT departments evaporates as corporations enter the New Normal. In the first leg of the digital revolution,

IT had a strong power base, as business executives depended on the ability of IT to guide them through the wonderful world of technology. But in the second leg, that power base is gone. The choice for IT departments is to either melt away, or adapt and reshape to reflect the new realm.

Today we still have armies of trained 'builders' in our IT departments, people we absolutely needed in the first leg to construct our IT systems. They are like the myriads that were necessary to build the pyramids, but became idle once the pyramids were built. We built IT systems to last forever, or at least were designed that way, but we now need IT to help businesses be flexible, agile and capable of turning on a dime.

This pressure on IT is not a transition phase. It is here to stay. It is the New Normal of IT.

THE FOUR I'S

Clever businesses will anticipate the necessary changes, and will dismantle the old IT and transform the way we think about technological innovation. In this book we will talk about the four I's that are the pillars of a re-imagined technology paradigm:

❶ INFORMATION

Yes. IT stands for Information Technology, but in the first half we focused on the 'T', while in the second half we will have to focus on the 'I': Information.

As John Naisbitt wrote in his great book 'Megatrends' in 1982: "We will be drowning in information, and starved of knowledge." He was right. Today companies are overflowing with emails and documents, flooded with templates and folders, and inundated with intranets, yet nobody finds anything. While we have certainly become more efficient producers of content, do we use that new content to increase our overall productivity? In the first leg we focused on building a technological foundation to *enable* knowledge workers to collaborate, but now we have to shift and build the skills, competencies and attitude to be more productive with that information.

John Naisbitt, *Megatrends.Ten New Directions Transforming Our Lives*, 1982

In a special report called 'The Data Deluge' by The Economist (25 February 2010), a single graph illustrates the information overload and the growing gap between storage capacity and the amount of information being created.

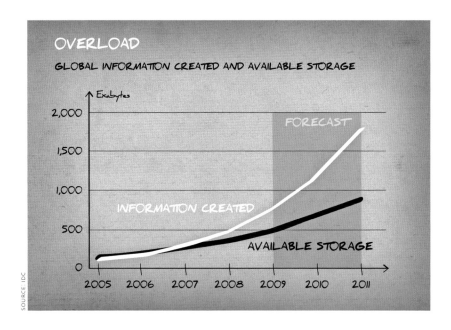

❷ INTELLIGENCE

How many databases does your company have? How much data do you handle? How accurate is your data? How much is data and how much is real intelligence? We store zillions of datapoints, and the overall reaction in most boardrooms must be: "Nah. The numbers are wrong." We have more data than ever before, but we just can't seem to turn them into intelligence. We built databases so fast that we can't see the forest for the trees. Then we built data-warehouses *on top* of the databases to make sense of the data. Then we decided not to trust those either. We've piled system on top of system, spent billions on capacity to *store* data, but that no longer works in the second half. The second leg means a focus on *intelligence*, not on data storage.

❸ INTEGRATION

A company isn't an island anymore. We've spent the last ten years building connections between systems *within* our organizations, but we'll spend the next ten years integrating our information with information *outside* our organizational boundaries. The skill of the second journey is the capability to cleverly connect, while the essential skill of the past was to masterfully build.

The most important of the four I's is the mindset we need to adopt. It's not about technology anymore: it's about being *clever* with technology. It's about technology-enabled innovation. When access to technology becomes a commodity, we no longer differentiate by just using technology. What we need to do is to develop the capacity to innovate *with* technology.

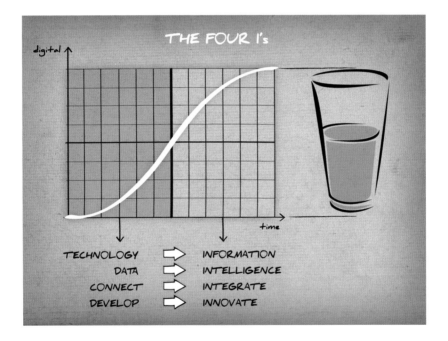

THE FOUR I's

TECHNOLOGY	⇒	INFORMATION
DATA	⇒	INTELLIGENCE
CONNECT	⇒	INTEGRATE
DEVELOP	⇒	INNOVATE

This is quite a quantum jump. That's why the transition will be hardest on the old IT department. It's about making that change, kicking off that first stage of the rocket and burning up the boosters in the second stage to transcend the earth's atmosphere.

BITS AND ATOMS

Some companies will be affected more than others. In his groundbreaking book 'Being Digital' in 1995, Nicholas Negroponte framed the difference between 'moving bits and moving atoms'. Some companies are almost purely 'bits' companies, like banks and insurance companies. Their real assets are nothing more than 'bits' of information stored and transported. Conversely, some companies are 'atom' companies, profiting from the movement of real physical material. The core question is:

Nicholas Negroponte,
Being Digital, 1995

Can the product or service that your company offers be digitized?

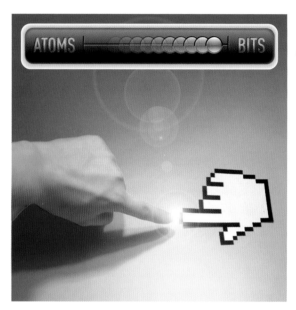

What we've observed since Negroponte's book is that some 'bits markets' have started feeling the heat. Just look at the massive changes we've seen in the media business:[1] print media has lost classified ads to online variants like Craigslist, ad revenue on television is declining rapidly as more eyes view online, and the music industry has been decimated by online offerings, first illegally on Napster and now legally on iTunes.

The pure 'atoms markets' have been impacted much less significantly. If you run a coal mine in Australia, or operate a quarry in Zimbabwe, the chances that you've been obliterated by the New Normal are much slimmer. Realistically, most companies are not 100% bits or 100% atoms but the 'Atoms to Bits ratio' is a good predictor of *when* the New Normal will hit you.

[1] The Economist:
'The year of the paywall'
Newspaper executives had assumed (or, perhaps more accurately, hoped) that advertising revenue would gradually migrate from print to the web, together with readers. The digital-revenue wobble threw that into question. Worse, newspapers are losing market share online, where advertising is moving away from banners and classifieds towards search.

Take the travel industry – there are plenty of atoms here: planes, hotel rooms, fuel, swimming pools, etc. But the pattern of interaction between travel companies and their customers has turned almost exclusively digital. The industry changed dramatically when the New Normal hit them. Today, almost every customer uses digital technology to find out information about a travel destination, using the Internet to compare different offerings and increasingly using digital means to reserve, book, pay and give feedback on their travels. The fundamental goal of the industry – to move travelers from one destination to another – is still very much intact, but the details have changed immensely.

When the New Normal hits a market like the travel industry, the impact can be devastating on the bit-centric elements of the business, but it impacts the traditional 'atoms' parts too. In the travel sector, the 'rhythm' used to be very simple: two major releases per year. The travel companies produced a winter catalogue and a summer catalogue and the whole organization revolved around those two major peaks of activity per year. Once the sector hit the New Normal, the rhythm of two releases per year evaporated and the operations of the companies became a 24/7, 365 days a year operation.

The companies in the travel sector that were able to cope with the New Normal were those who repositioned technology not as a 'side activity' in the IT department, but made technological innovation the *core* of their activities, leading them to innovate new ways to interact with their customers and partners. These changes aren't IT changes; they are business changes. It requires business attention and business oversight to make this happen.

CAN YOU ADAPT?

Can you adapt? Can you adapt to the New Normal, and can you do it quickly enough? Can you shift your mindset towards the realities of the New Normal?

Over time, the problem will disappear. In twenty years, most managers will be digital natives, and everyone else will have adjusted to the New Normal. But the *transition* phase between the halfway point, where digital becomes the New Normal in society, and the point where digital has become the New Normal in the boardroom, will be a crucial one.

It is your responsibility to manage your company through those challenging years beyond the halfway point. Your challenge is to navigate your organization through the storm of the New Normal.

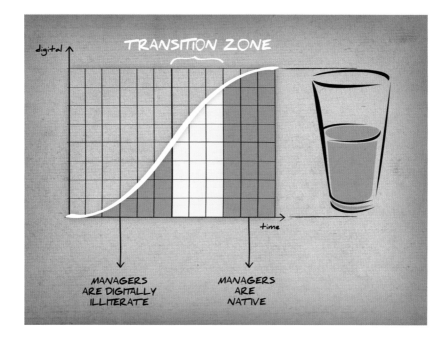

ONWARDS HO

In the rest of this book, we will look at the New Normal and explore its limits. We will discover what happens when we're fully submerged in the deep end of the pool. We'll explore the impact of the New Normal. We'll study what the limits will be when we look at attention span, intelligence, pricing, privacy and control when navigating the second leg of the digital journey.

We will also construct 'rules of the New Normal'. We will review the rules which will govern the second half of the game, and how we will have to behave as organizations to keep employees and customers on board in the New Normal.

Next we'll tackle the four I's. Information, intelligence, integration and innovation are the mantras of the New Normal. This book will clarify what they mean, how can you build a strategy in those areas, and how can you practically tackle the challenges ahead.

Finally, I want to reflect on IT itself. How can we reshape the old IT, and how can the business take ownership of technology-enabled innovation? What will be the 'shape under the drape' when we examine where technology is leading us?

You are at the crossroads of the digital revolution, and about to step into the great unknown. Our feet are feeling the curve of the pool dropping away below us, and we can sense the attraction of the deep end of the pool. We feel the tug of the New Normal all around us, in society, in our family, in our colleagues who are being drawn into the realm where being digital is just being normal.

EXPLORING
THE LIMITS

"The beginning is half of the whole."
— Pythagoras

"The future just ain't what it used to be."
— Yogi Berra

A brief refresher on limits. I promise this won't hurt.

A limit is a mathematical term that refers to 'what happens in the end'. Maybe you will remember a classic series of figures, such as this one:

$$\sum_{k=1}^{\infty} \frac{1}{k} = 1 + \frac{1}{2} + \frac{1}{3} + \frac{1}{4} + \frac{1}{5} + \frac{1}{6} + \ldots =$$

This series sums to infinity. If you keep adding up all these terms, the number you come up with keeps growing larger and larger, and it won't 'converge', but will actually go to infinity. So, we say that the 'limit' is infinite, and we write:

$$\sum_{k=1}^{\infty} \frac{1}{k} = 1 + \frac{1}{2} + \frac{1}{3} + \frac{1}{4} + \frac{1}{5} + \frac{1}{6} + \ldots = \infty$$

Other series of terms behave differently. For example:

$$\sum_{k=0}^{\infty} \frac{1}{2^k} = 1 + \frac{1}{2} + \frac{1}{4} + \frac{1}{8} + \frac{1}{16} + \frac{1}{32} + \ldots =$$

This series doesn't reach infinity. As it turns out, if you add this infinite set of terms together, you actually converge to a finite number. And that number is plain old two:

$$\sum_{k=0}^{\infty} \frac{1}{2^k} = 1 + \frac{1}{2} + \frac{1}{4} + \frac{1}{8} + \frac{1}{16} + \frac{1}{32} + \ldots = (2)$$

While the mathematical notation for series and limits is not necessary for our thought experiment in the upcoming sections, it will ① put you in the right frame of mind to consider where we are headed in the end, and ② impress your seat mate on the airplane.

THE LIMITS OF THE NEW NORMAL
WHERE TO GO IN THE 'END'?

As we've seen in the first chapter, the New Normal is the exciting world that we are entering in our personal lives, a world where we take digital for granted.

In the last 15 years we've heard about the disruptive capacity of digital on traditional business models, particularly with the first Internet hype in the late 1990s, and perhaps more interestingly with the resurrection of the Internet during the Web 2.0 phenomenon.

These cycles of digital reawakening within the business community are happening faster and with more intensity.

But I'm getting tired of the 2.0 thing, and even wearier of the 'New New Thing' that everyone is chasing. We only started getting used to people wielding the 2.0 term when the smart and savvy among us started asking when the 3.0 thing would pan out.

We don't want to look at 3.0, or 2.5, or even 4.0.
What we want to do here, is take it to the limit. All the way.

I want to explore some functionalities of our digital behavior, once we've passed the halfway mark. I want to observe our digital interactions once we're clearly in the New Normal zone. But rather than look at incremental step-by-step evolution, let's jump all the way to the sum of the series – let's explore the limit and discover where it will go in the end.

I don't know when we will reach these limits. They're positioned in some hypothetical future. But most importantly we're going on a mental journey, a *'Gedankenexperiment'* that will take us far into the New Normal, and allow us to reflect on where we might end up. Let's go.

THE LIMIT OF LENGTH

In the 'end', the amount of information that we ingest will inevitably converge to zero. My statement is that the limit of length goes to zero.

$$\text{LIMIT (LENGTH)} = 0$$

Let me explain.

To graduate from university, I was required to write a **thesis**. This was probably the hardest project I had to do in my life. The task of writing a coherent piece of information with, hopefully, some new insights into a particular subject, at least 100 pages in length, was in my eyes a daunting task of biblical proportions.

Although you knew your subject a year in advance, the typical approach to writing a thesis was to put off writing any actual text for as long as you could. You would do research, read books, articles and papers on the subject, gather a truckload of information that you piled on your desk, all the while approaching your deadline. As my favorite author Douglas Adams said: "I love deadlines. I especially love the swooshing sound they make as they go flying by."

www.neonormal.com
Take it to the limit

Eventually, I finished my thesis with the smallest time margin possible, having to simultaneously print the document on four parallel laser printers to get it on the desk of my professor in time.

It was 132 pages long, and I got a mark of 17 out of 20.

In my university, there was an absolute correlation between the number of pages of your thesis and your final mark. In fact, from the pedagogical standpoint, the express purpose of writing a thesis was to check if you *could* actually write a coherent document of that length, and consequently you got rewarded based on how much paper you produced.

When I got my first job after graduation, I was still in thesis mode: study a particular subject, research it, and write a long paper about it. To my amazement, everyone around me was busy writing a memo.

A **memo** is quite different. A memo is a mini-document. A memo is meant to be precise, and, more importantly, concise.

My first memos were disasters, and within a few months my boss sat me down and said: "We know you're smart. That's why we hired you. But don't show off on every damned memo. You're not writing a thesis anymore; write what people will actually read."

From then on, I understood how corporate life worked. I had to produce information not based on the 'source of', but I had to create information based on the 'consumption of'.

It is not about the sender. It is about the receiver.

My memos got shorter and shorter. In a few years time I had it down to an art form, writing single pages which were read by everyone.

And then **email** erupted on the scene and transformed it all over again.

Suddenly, our memos could be sent to everyone with the greatest ease, plus commented on and resent. The longer documents we still had but no one read, were now packaged into attachments that we could send around the company to show how incredibly smart we had been to write all of this. No one read these attachments, but at least we could broadcast them to the entire office.

People found a great amount of really neat content online – whitepapers and brilliant position papers, and super thought-provoking Power-Points and PDFs – and we all sent attachments to each other with a little note, and we all ignored each other's attachments.

The results have been dramatic. We have an overload of information, have a gazillion channels attacking us with information on a daily basis, and our intake of information actually gets smaller and smaller.

The irony of our situation is that although we've never had more information at our disposal, we seem to absorb less and less.

Recently, I was asked to work on an interesting consulting assignment at a leading software vendor: to quantify how their employees use productivity tools. Essentially, I had to walk around the offices, observe the employees in action with their *own* tools, and provide a blunt critique of how their crowd 'ate their own dog food'.

On the first day, I met with the manager in charge of the assignment, and asked "So, what do you guys want out of this in the end? A report?"

He looked at me cynically: "A report? Do you think we have time to read reports?" "Perhaps a PowerPoint presentation then?" I tried.

His answer was quite revealing. "Oh please no. God no." After a moment of reflection, he proposed that I write a column: "But make it short, and make sure it has a snappy title or no one here will ever read it."

There I was, in the offices of one of the leading companies in the world that had transformed the way we handle information, that had singlehandedly transmuted our offices from paper-based to digital, and he put his finger on a very interesting point:

"People don't ingest information anymore. They just push it around."

THE INFORMATION DILEMMA

Digitization has made it so simple to search, retrieve, store, mail and forward information, that we have gotten ourselves into a very strange situation.

The amount of information we get 'hit' with on a daily basis is staggering. The number of channels that 'aim' at us has exploded, yet the amount of information that we really 'ingest' is shrinking.

The problem isn't information overload, but filter failure.

Email is an interesting example. When email first came out, I wanted more. I was actually asking my friends to 'drop me an email', probably because I was a little worried that the email system I was using wasn't always working correctly. I actually begged people to write me an email out of insecurity.

Today I get 150+ emails a day. It has become a horrible hassle. The overload gets so bad that I *have* to take my Blackberry with me on a vacation, because the amount of work I need to do to catch up when I come back takes all the fun out of the holiday experience.

In the early days of email, I tried to go home every night with an inbox that was 'survey-able'. It meant that I didn't want to leave the office if I still had to scroll in my inbox. I would work and work and work until I finally could no longer see a scrollbar, and then it was time to go home. I bought bigger and bigger monitors, until finally I gave up and accepted the fact that I could no longer survey my inbox with just one glance.

I've met people a lot worse than me. At every lecture I give, I try to find the poor sap in the audience that gets the *most* emails. He or she is typically not the happiest person in the crowd either. Remember just a few years ago when your corporate importance was rated by how *much* information you got?

Now if you meet someone who only gets five emails a day, you know this person is really important, because the filtering around him is so great that he *only* receives the most important emails.

It's all about filtering.

Let's get back to the limit. The limit of length goes to zero.

Last year I started using **Twitter**. I just had to understand how it worked. Again here I had my doubts about the medium. Why would I want to know what my neighbor had for breakfast or how someone is doing while giving birth. But it didn't take long before I became to adore the tool.

The length of a message in Twitter is limited to 140 characters. Max. That is the amount of information that we are able to absorb today. But if we extrapolate Tweets to the limit of the New Normal, the limit of length goes to zero. Nada.

THE LIMIT OF DEPTH

The limit of depth, on the other hand, goes to infinity.

$$\text{LIMIT (DEPTH)} = \infty$$

Although the 'attention span' of the average user goes to zero when unwanted information is forced on him, the capacity for the average user to access information that he desires is limitless. When we actively pursue information, the possibilities to seek have become infinite.

There are plenty of subjects I know very little about, but Chinese Pottery is something that I know absolutely *nothing* about. However, if you give me 15 minutes on Wikipedia, I will be able to have a decent conversation on the subject. If you give me an hour on the Internet, I can probably give a lecture on Chinese Pottery. With illustrations.

Our possibilities to research and *find* the right information have become endless.

I recently spent time researching healthcare and pharmaceutical companies. The game changing effect of the New Normal is very interesting to observe in this area. For a very long time, the role of the doctor was pivotal. The patient went to the doctor, got an examination, and received a prescription. No dispute.

www.neonormal.com

Today, patients have examined their symptoms online long before they step foot into a doctor's office, referencing sites such as 'www.patientslikeme.com' to see what diagnoses other patients received, plus what treatments, and what dosages.

Today when a patient walks into the doctor's office, he is more likely seeking a prescription than a diagnosis. Here, the value of the doctor becomes his legal status, not his knowledge base, which has largely been digitally diffused at a general practitioner level.

Knowledge workers in the health industry, like knowledge workers in almost all sectors, will continue to have their authority eroded by the drill-down possibilities of a digital world, though they will likely cling to their legal authority long after the knowledge gap has closed.

The physician's relationship to the pharmaceutical companies is also evolving. The traditional model for pharmaceutical company sales was to employ armies of 'medical representatives' who would patiently sit in the waiting rooms of the doctors and explain to the physician the benefits of a certain type of treatment.

No more. The time allotted to the sales reps is declining, and will evolve to zero. The limit of the 'length' of time a sales rep has with the doctor goes to absolute zero. But *if* the physician wants to find out anything about a particular drug, his possibility to investigate it himself goes to infinity. The limit of depth is infinite.

We see this pattern in every single area in the New Normal. Consumers have less and less attention span, but when they *want* information, their possibilities to research, zoom in and seek have become endless.

Within your company, employees are overloaded with information and channels, and the attention span goes to zero, but when they *want* to unearth the right information, you will have to ensure that your information systems allow them to go to infinity.

THE LIMIT OF PRICE

As the New Normal continues to push the world's 'Atoms to Bits ratio' toward bits, what is the limit of price? Chris Anderson, editor of Wired Magazine and author of the acclaimed 'The Long Tail', recently wrote an interesting book on this topic. His book, 'Free: The Future of a Radical Price', argues that the limit of Price moves to $0.00.

Chris Anderson,
The Long Tail: how endless choice is creating unlimited demand, 2006

Chris Anderson,
Free: The Future of a Radical Price, 2009

$$\text{LIMIT}\,(\text{PRICE}) = \$0.00$$

At first glance, Stewart Brand's famous quote "Information wants to be free" points us in the same direction, though this snippet is taken out of context from his full statement in his book 'The Media Lab: Inventing the Future at MIT': "Information wants to be free. Information also wants to be expensive. Information wants to be free because it has become so cheap to distribute, copy, and recombine – too cheap to meter. It wants to be expensive because it can be immeasurably valuable to the recipient. That tension will not go away. It leads to endless wrenching debate about price, copyright, intellectual property, the moral rightness of casual distribution, because each round of new devices makes the tension worse, not better."

Stewart Brand,
The Media Lab: Inventing the Future at MIT, 1987

There are strong examples on both sides of the argument. In the realm of Anderson's 'Free', we embrace free digital services like Google's Gmail, Google Docs and Picasa. We also enjoy free digital content from newspapers, magazines, and highly trafficked blogs like Boing Boing.

On the flip side, as Malcolm Gladwell noted in his critique of 'Free' in the 'New Yorker', drug companies continue to make tremendous amounts of money from intellectual property that is not trending toward free. As for Apple, which makes money on both storage (iPod) and content (iTunes), it is still unclear which piece will become the razor and which piece will be the razor blade.

www.newyorker.com
search on:
author: Gladwell
date: 06/07/2009

In the Apple example, we see an interesting interplay between bits and atoms. The content of an 'iTune' are bits, which have been created by a musician. Technically, Apple can reproduce these bits one million times for virtually nothing. Legally, any artist selling music through iTunes is entitled to a percentage of the purchase price of each song (typically 66% of $1.29), so it costs Apple more to sell one million songs than to sell one song, though of course their revenues exceed their costs and their profit margins are phenomenal.

The storage device –the iPod– is made of atoms, though as we know from 'Moore's Law', the costs of some component pieces of the iPod are decreasing while performance is increasing.

Choose a limit

LENGTH   DEPTH   PRIVACY   PRICE   INTELLIGENCE   PATIENCE

DIGITAL   WE ARE HERE

2ⁿᵈ leg

$0.0

TIME

-25 YEARS   NOW   +25 YEARS
                          LIMIT(PRICE)=0.0$

www.neonormal.com

At the limit of the New Normal, where will Apple make its money? There are already cheap storage devices that can play audio files with the same quality as the iPod, and yet the storage device market has not been commoditized. Here we see that Apple's design (i.e. intellectual property) continues to drive its success on the hardware piece. Will Apple's intellectual property/hardware move toward Free, leading them to give away iPods and concentrate on distribution of iTunes?

Conversely, there are plenty of ways (not necessarily legal) to load your iPod with free music. Will Apple use its leverage over content producers to drive the price of content toward Free, while profiting off of its storage devices? In either case, it seems that some bits will continue to be valued. So the limit of price is... unknown.

THE LIMIT OF PATIENCE

As the attention span of users has dropped dramatically, so too has their patience for digital redundancy. A user understands the virtually unlimited capacity of digital brawn (storage), and now demands a matching level of digital brain (intelligent database search).

Said differently, users hate providing info more than once, because if the digital system has seen it once, the digital system should be able to find it again. Therefore, my statement is that the limit of patience goes to one; users want to provide information *once* and *once only*.

LIMIT (PATIENCE) = 1

Some companies have been ahead of the curve in predicting the limit of patience and positioning their products accordingly. In the late 1990s, Pay-Pal sensed growing consumer frustration in the nascent but fragmented e-commerce landscape, and built a billion dollar business on the twin foundations of convenience and security.

The strategy has paid off: consumers embrace PayPal as a quick and easy e-payment instrument, and in 2009 15 percent of all global e-commerce was conducted using a PayPal account.

The limit of patience manifests itself in several other ways as well – in the midst of abundance, consumers want less choice and less (visible) complexity.

Barry Schwartz' bestselling book 'The Paradox of Choice: Why More Is Less' lays out a convincing argument for the pitfalls of overwhelming consumers with choice. In a simple example Schwarz reviews an experiment where shoppers encounter a seller offering six

Barry Schwartz, *The Paradox of Choice: Why More Is Less*, 2004 types of jam, contrasted with the same scenario involving twenty-four types of jam, including the original six. Interestingly, the shoppers are more likely to purchase jam when confronted with a simpler menu of options.

Technology has definitely played a big role in this explosion of choice by, among other things, increasing the efficiency of supply chains. However, it can also play a role in creating simplicity out of abundance. Companies that understand the limit of patience and the power of the filtering have succeeded in the marketplace.

Google is an obvious success story, but Amazon.com is perhaps the best example of a company that balances a massively large number of choices with a filtration system – powered both by algorithms and crowd-sourcing – that allows its customers to reach a 'buy' decision.

As for a reduction in visible complexity, part of the reason why Apple was able to capture such a huge portion of the market with the iPod and the iPhone was because they built on the philosophy that is embedded in the design of the Macintosh; hide the complexity of the system from the user, and make sure that you optimize the efficiency of the end user.

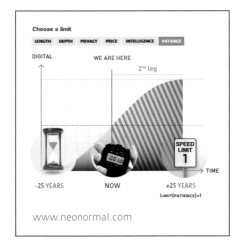

www.neonormal.com

Microsoft-based systems are a delight for the engineer and sometimes a challenge for the end user, and Apple products are the other way around: often a nightmare for the engineers to program the machines, but a fun experience for the end users. Anyone who remembers the Microsoft Vista debacle knows that the patience of users is very low, and will drop to 1 in the New Normal.

One of the most interesting limits is the limit of privacy. This is where a true generational aspect comes into play. Privacy is one of the areas where you will recognize the impact of the New Normal, because when we cross the halfway mark, our notion of privacy becomes a notion of transparency, and in the limit this evolves to a 'fishbowl society', marked by total transparency.

LIMIT (PRIVACY) =

Take the example of recruiting. Engaging in relationship due diligence (Googling or Binging) has become common practice with our business partners and customers, but generally these folks are digital immigrants who are more likely to hold on to a traditional view of privacy in the realm of sharing personal info on the web.

Our new entry-level recruits, on the other hand, embrace a view of privacy that can point us toward the New Normal limit. Before a face-to-face interview with a new recruit, we will definitely spend a few minutes on their LinkedIn profile, but the revealing bits come out when we look at their personal pages on Facebook or MySpace, or their feeds on Twitter.

Can we make a value judgment on this shift in attitudes toward privacy? On one hand, on these personal pages, we can find embarrassing pictures that don't always put the recruits in the best light. Do we need to see our prospective junior analyst's victorious beer pong game from last Saturday night? Conversely, there are also unexpected points of connection – a shared love of Roman history or table tennis – that might help form a lasting impression on the recruiter.

Regardless of how different generations may view the appropriateness of this online glut of personal information, this trend toward openness will not abate. Once we cross into the New Normal, privacy becomes a thing of the past. Our experiences online have played a huge part in this transition.

In the limit of patience, we saw the expectation that the digital world should always be able to recall useful information. Here we see that the digital world also recalls information that we wish it would forget – it is hard to 'erase' something online. I'm still amazed when I Google my own name to see how many quotes, articles and documents surface that I had long forgotten, but that the Internet has not.

I'm even more amazed to read some of the things I've said, or even worse predicted, and see how foolish or wrong I've been. And I'm especially distressed to see my old pictures and YouTube videos of the past being played out. But there is very little I can do about it.

There are today a number of specialized companies that make a living by 'erasing things' online. You can pay them a hefty sum of money to have some elements of your online past taken out, but this reminds me very much of the shady deals where you fork over a large sum of money to some monosyllabic thug to 'take care of a little problem'. The truth is, erasing something online is extremely difficult. Hiding something online has become nearly impossible as well.

So it is no surprise to see the reaction of Generation Y when they cross to the New Normal and adopt an attitude of: "Well. We might as well get it all out in the open anyway, because it is pointless to try and hide." I applaud this mentality.

One of the most popular iPhone apps is the application to check out a potential suitor, on the fly. You start up a casual conversation in a bar, and then while that person is in the bathroom, you take out your iPhone, enter the name and details of your potential partner, and instantly have all known information on your date. Married or not, previous marriages, previous children. Is he paying his alimony on time? Any criminal records? Any outstanding fines? Wanted in any particular state? Plenty of topics to talk about when your date returns from the bathroom. The prevailing attitude is one of complete access to all relevant information. Total transparency.

We'll have to learn to live with this attitude, and we will have a very interesting transition period finding our way, in part because of the generational divide. Almost every country has some sort of a 'privacy commission' that rules on what is appropriate behavior in terms of privacy, and they tackle some pretty serious subjects.

Recently I talked to one of the junior members of the privacy commission of a key EU member state, and he complained that the 'honorable members of the privacy commission' in his country were, on average, seventy years old, and had probably never touched a computer in their lives. Yet they are nonetheless being tasked to rule on pressing electronic and digital privacy boundaries for the Facebook and Google generation. I could sense his frustration.

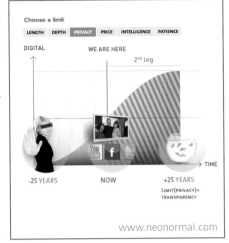

www.neonormal.com

The situation in companies is equally complicated. Boundaries of companies are blurring. Employees have gone from long-term engagements towards shorter-term, project-oriented interactions. How do we manage and secure information in a world of complete transparency and connectedness?

We're living more and more in a fishbowl society.

Transparency is the norm, not the exception.

The limit of privacy will be one of the most difficult ones to tackle.

THE LIMIT OF INTELLIGENCE

The final limit I'd like to talk about is the limit of intelligence. My statement is that in the New Normal, the limit of intelligence becomes a 'real time' issue:

LIMIT (INTELLIGENCE) = REAL TIME

If you look at the history of technology, and in particular Information Technology, what we really wanted with this computing power was to be able to predict more accurately. Computers were used for the first time during the Second World War to predict ballistic trajectories of shells and rockets, and this technology was then widely adapted to a number of industries after the war.

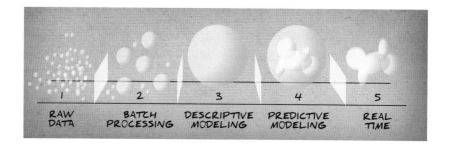

1	2	3	4	5
RAW DATA	BATCH PROCESSING	DESCRIPTIVE MODELING	PREDICTIVE MODELING	REAL TIME

Banks started using computers to predict the markets, insurance companies to foresee the evolution of risk and premiums, meteorological institutes to forecast the weather, and retail chains to observe the behavior of customers and pick up trends faster than competitors.

But once we had systems working through piles of numbers to reveal intelligence, and once we started processing large batches of this data, we saw the need to speed up the intelligence towards faster and faster cycles.

We evolved from batch processing –nights of working with the numbers– towards modeling techniques, and now increasingly towards almost 'instant' use of intelligence and information.

Let's give an example. In the first Internet hype, when companies started building their first websites, we liked to examine the logfiles of our brand-new site. We would pore over the logs and see who had visited us online, where they were from, what they did, and which pages they consulted. We counted how many 'hits' we had on our site.

'Hits' now stands for 'How Idiots Track Success'.

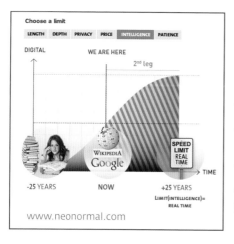

But once we've crossed into the New Normal, we need to act faster and faster. We don't have time to analyze the log-files of our visitors or to process data after the fact. What we need to do is instantly move and react. 'Hits' are in the past tense. We need to know who is in the process of 'hitting' the site, why they are 'hitting' the site, and how to turn their surfing behavior into buying behavior.

The moment that a customer, prospect or potential client visits your website, you need to act. Tools like whos.amung.us help us monitor our web traffic in real time. You must capture that moment and react with all the knowledge and history you have on that customer.

Services like Salesforce.com's real-time business analytics help to make sense of up-to-the-second data. You have to make the site as relevant as possible, provide the most tailored information, and 'capture' the visitor in real time.

The limit of intelligence requires absolute intelligence, right now. Business in the New Normal happens in 'real time'.

SO, WHAT ARE YOUR LIMITS?

This list of limits is by no means complete. We encourage people to explore other limits, and other extremes that will occur in the New Normal.

But it is probably a good idea to take some time to reflect on these limits for your company, your organization and your context. What happens when you cross into the New Normal? What will this mean for your employees, for your customers and your markets? What will happen not just next year, or the year after, but in the 'limit'? Where will things go in the 'end'?

Wayne Gretzky, the retired Canadian professional ice hockey icon, said: "A good hockey player plays where the puck is. A great hockey player plays where the puck is going to be."

That is true of successful people in all fields. We have to look further into the future rather than react to 2.0, and then 2.5, and then 3.0... As we saw in this chapter, companies from Google to PayPal to Amazon have exemplified the advantage that can be gained by thinking ahead of the curve, and positioning products accordingly. In today's ever-changing environment, it pays to look at the limit of where we are going, because we are getting there fast.

Take a moment to think about your limits. The next step will be to deduce some new 'rules' for the New Normal.

First the limits. Then the rules.

CHAPTER 03

RULES OF
THE NEW NORMAL

"Essentially, all models are wrong,
but some are useful."
— George Box

"Implementing best practice is
replicating yesterday. Innovation is
designing tomorrow."
— Paul Sloane

"How do you recognize something
that is still technology? A good clue is
if it comes with a manual."
— Douglas Adams

In the previous chapters, we defined the concept of the New Normal, where digital becomes the standard, and observed what happens when we look at the extremes of user behavior in the 'limit' of this New Normal.

In this chapter, we want to construct some 'rules of the New Normal' to guide us through our journey once we've passed the digital halfway mark.

 ZERO TOLERANCE FOR DIGITAL FAILURE

Over the last few years, society has become neurotic about digital failure. This will only get worse.

The first rule of the New Normal is that there will be 'zero tolerance for digital failure'. This rule dovetails with our limit of patience: users understand the capacity and perfectibility of the digital world, and accordingly demand that digital offerings be 100% reliable. It also connects to the ubiquity of digital in the New Normal.

An anecdotal example can be found in my own home. A decade ago, when our Internet access would fail for whatever reason, I was the only one who was inconvenienced. This was the era when digital technology was primarily a work tool, so while I labored to regain access to my emails, the rest of my family barely noticed the outage.

Today is a different story. If our Internet connection fails, the daily routine of each member of my family is severely disrupted. We can't access weather forecasts, recipes, social networking sites, driving directions, etc.

Without digital, we can barely function. So digital has to be available and working all the time. No excuses. Always on.

I recently visited a government agency that I had been to many times before, but this time something was different. I couldn't put my finger on it, but I noticed something out of the ordinary the moment I walked in the building. There was a lot of activity: busy corridors, crowds of people around the water coolers and coffee machines, a great buzz of chatter all over the building.

When we sat down in the conference room, my host told me the problem: a network failure had completely wiped out all connectivity and

rendered the organization completely immobile. No mail, no Internet, not even access to any documents or PowerPoints. But what great cohesion and bonding it generated in the hallways!

A digital failure – defined here as an inability to move and manipulate bits – is now akin to an unexpected snowstorm that temporarily impedes our ability to move atoms. Imagine a snow day, where offices and schools are shut down and we experience bonding around this unusual event.

The more we move into the New Normal, the more we become digital, the more we depend on digital, and the more we can't afford for digital to fail.

In 2009, Gmail had two major outages: one for four hours in February, another for two hours in September. These were, in fact, the first real outages of Gmail, ever. Tens of millions of people had come to rely on Gmail as their email system of choice, and had become very used to the fact that Gmail *always* worked.

For two full hours in September, millions of users could not access their email from Google. It made CNN headline news. The media had a field day, the Internet was buzzing, and users were outraged because they could not check their email for those two hours. It was a shame.

You must acknowledge that Gmail had been 'up' for almost five years. For five consecutive years, Gmail had been running 24 hours a day, 7 days a week, 52 weeks a year. Without failure.

I know plenty of CIOs who'd be drinking champagne every single night in the datacenter if they had systems that only went down for six hours every five years. But in the New Normal, that behavior is not acceptable. In the New Normal, there is a zero tolerance for digital failure.

FAMOUS INTERNET FAILURES

Over the years, the Internet has failed us a number of times. Some famous examples:

1 DAY

In **JUNE 1998**, the eBay auction site was out for almost a full business day. eBay reportedly lost between three and five million dollars on that single day and was forced to offer refunds, extend auctions and waive fees as an apology to its customers.

2 DAYS

In **SEPTEMBER 2003**, an Internet worm wreaked havoc in the United States. The Gibe-F worm, an email-transmitted virus, initiated cascading server failures. Within an hour, Internet service to more than 90 percent of the U.S. was disabled, either by the worm or by network firewalls that initiated security protocols. In many companies, personnel was still able to send mail internally, but for a full 48 hours most knowledge workers in the U.S. were completely cut off from external email and Internet services.

2 DAYS

In **AUGUST 2007**, the telephone landline alternative Skype went missing for two days. The outage was blamed on an algorithm within the Skype networking software. Skype recovered from the incident quite easily, mainly because the service blogged openly about the failure.

1 DAY

In **JANUARY 2008**, the Internet went down in large sections of the Middle East, Asia and North Africa. India lost half of its broadband connectivity. This caused a good deal of trouble for the outsourcing industry in India, which relies heavily on broadband communications with its customers in the rest of the world. United Arab Emirates also reported problems. Dubai is well known for its billion dollar a day transactions, which were impossible on January 31st 2008. The Internet outage was blamed on a cut in an undersea cable.

Digital failure in the New Normal is more than just inconvenient: it is scary. In 2009 and 2010, we experienced a massive Toyota car recall because of a problem with 'unwanted acceleration' in some of their models. One of the big questions was whether the problems in the Toyota vehicles were caused by a mechanical problem or an electronic problem. Any failure is, of course, unacceptable, but it became clear in the public debates that mechanical failures were something that you could fix (which Toyota actually did by recalling the vehicles and adding a strip of metal to the accelerator pedal).

If the source of the problem had been electronic, this would have caused much greater distress in the general public. Indeed, there is a comforting physicality to mechanical problems. You can 'inspect' an extra piece of metal on your pedal ("That looks better. That will do the trick."), whereas digital modifications take place within a black box, inscrutable to the end user. Has the problem actually been fixed? You can only hope so.

We can imagine the impact of digital failures on a much larger scale, though thankfully we can't point to any examples of massive outages beyond dystopic sci-fi novels and Hollywood thrillers. Digital technology powers L.A.'s sprawling traffic grid, coordinates the landing schedule at Heathrow, keeps Dick Cheney's pacemaker ticking, etc. We increasingly depend on digital without a manual override, and we will only continue on in that direction.

No tolerance, not any more. Zero tolerance for digital failure.

GOOD ENOUGH BEATS PERFECT

There is a subtle distinction in the newly emerging digital world between 100% accessibility (as dictated in Rule #1) and 100% perfection. In fact, as we move into the New Normal, speed, ease, and availability become more important than perfection. Velocity trumps perfection in the New Normal, or in other words:

> Inspired by Robert Capps, *The Good Enough Revolution: When Cheap and Simple Is Just Fine*, Wired magazine, 24 August 2009

Welcome to the era of good enough technology.

We (in IT) used to strive for perfection. We tried to build the best there could possibly be. We aimed for the sky and beyond. We built Rolls Royces. Sometimes what we built didn't work properly, but we *aimed* for perfection.

Once we cross the New Normal, that mentality is obsolete. No more tolerance for failure, and above all, make sure it is fast. Flexibility is more important than the ideal.

The perfect is the enemy of the good, especially if the good is cheaper, faster, or more convenient.

We see this type of good enough technology in our everyday lives. As Robert Capps from Wired Magazine noted: "Cheap, fast, simple tools are suddenly everywhere."

Let's look at a couple of examples.

Skype is good enough technology. I still prefer the phone and actually think the phone is better than Skype. It still gives me the best experience in terms of consistent sound quality. I don't want Skype to fail – that would violate the first rule of the New Normal – but for most practical purposes Skype is good enough.

Skype is fast, easy to use, convenient, and pretty much free. So Skype has proximity with a lot of the limits of the New Normal that we talked about in the previous chapter.

Another example is **MP3**. I remember when digital music first emerged, and when I bought my first Audio Compact Disk (it was 'The Police – Greatest Hits'). Most people born before 1978 probably remember their first CD, and I still recall that day I bought it and took it out its plastic holder like it was made from Plutonium, terrified I would get a scratch on it. When I see my kids throwing their discs around like they were frisbees, I understand in my gut the pace of technological innovation.

The Compact Disk was invented by Philips engineers who worried about the sampling of the music. Philips eventually settled on sampling at 44Khz, so they could fit exactly 72 minutes of music onto a Compact Disk. Their main worry was that the quality of the digital music would be inferior to that of the old analog music recordings, and that audiophiles would be put off by the inferior quality.

Years later MP3 came onto the scene. Instead of sampling at 44Khz, the initial MP3s had a much worse sampling quality, but you could fit a lot more music onto a disc or store more songs on your MP3 player or iPod.

MP3 is, in most cases, good enough technology. We don't really *need* all that absurdly high quality of digital recording. For most practical purposes, when we listen to music in our cars, on a train, or while we're jogging, we're not going to notice the subtle audio nuances in a Brahms concerto, or in a Brian Eno composition.

Sometimes the 'good enough version' is actually better. Take **Gmail** from Google. When Gmail came to life in 2004, Google wanted to provide an online version of email very much like Hotmail, which had been acquired by Microsoft. Hotmail was an instant success, and Microsoft paid a hefty price for the acquisition, but actually providing email via a web browser wasn't technically that difficult.

As a matter of fact, it was quite easy. Gmail from Google was initially very simple in terms of functionality, but it made a great promise to its users: if you use Gmail, you will never run out of capacity. Ever. We, Google, will pledge that we will keep upgrading our systems and storage, so that you never have to clear out your inbox ever again.

Actually, it was announced on the 1st of April 2004, and many thought it was a classic Google April Fool's prank.

In other words, Gmail made the promise for 'infinite' email capacity. And that was the big attraction for users of the Gmail system. Chris Anderson once said: " 'Your inbox is full'. What was that all about?" Gmail started out as a good enough technology, but quickly added new features to make the good enough technology even better than the regular version.

INFINITY MEETS LIMITS

We can digress for a moment to make a few observations on the history of Gmail, which relates to our overall argument about the pace and depth of technological change. At the time Gmail was released, it offered 1 GB storage compared to around 4-8 MB from Hotmail. Can you imagine that only six years ago, the limit on a Hotmail account was the size of a few songs or a handful of picture attachments?

Google aggressively got out ahead of the curve by offering a free email service not with double or triple the storage space of their competitor, but an exponential increase that attracted media buzz and instant consumer attention.

But, more interestingly, here is an example of even Google not thinking far enough into the future, to the 'limit of file sizes'. Although Gmail's storage constantly grew as Google added more servers to their system, they could not keep up with increases in consumer usage and file sizes. Introduced just two years earlier with the slogan 'never delete another email', in 2006 Google added the 'delete' button for the first time, and in 2008 web postings started showing up about people running out of room.

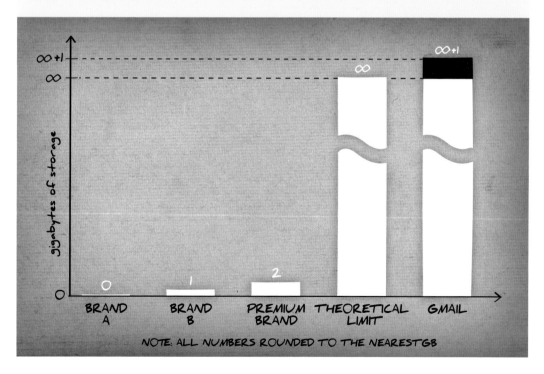

Back to good enough technology, where focusing on the end user is different than focusing on better technology (at least better defined in the classical measures of pixels, processing speed, etc.). A few years ago, the electronics industry was completely surprised by the success of **Net-Books**: inexpensive computers with little capacity and power, but that were neat, nifty and convenient. Users didn't always want bigger and bigger or meaner and meaner machines. What they really wanted was simplicity, portability and mobility. The NetBooks were good enough technology.

A final example. The **Blu-Ray disc**. One of the disappointments of the home electronics industry is the fact that Blu-Ray didn't take off as anticipated. The reason is quite simple: for most people, the DVD is good enough.

Unless you have a wall-sized plasma screen you're not going to notice the visual differences, and for the superior sound quality to really become apparent, you have to buy some pretty expensive speakers first. In my case, I've already bought the entire Disney collection from Bambi to Pinocchio on VHS *and* on DVD, and I'm not really inclined to buy them all again on Blu-Ray.

We're entering the 'era of good enough technology'. We don't want technology to fail, but it doesn't have to be perfect all the time.

Velocity trumps perfection in the New Normal.

THE ERA OF TOTAL ACCOUNTABILITY

The third rule is a peculiar one. It is about accountability. In my opinion, the New Normal will be marked by a sense of total accountability, primarily as a result of total transparency.

Let's take the example of the advertising agencies. One of my favorite television series over the last few years has been 'Mad Men', which gives a wonderful sense of the old-style Madison Avenue advertising world in the 1950s. As the show documents, these were the halcyon years of a booming advertising industry, a time when asymmetric information allowed advertising agencies to skillfully manage their customers.

The relationship between Blue Chip companies and ad executives has been an interesting one to observe over the years. Traditionally, the agencies would 'pitch' for a client, putting their skills in creativity and copywriting on display, and the customer would select an agency to handle all their advertising. The big buzz was 'Who got the Coca-Cola account, and who got the Nike deal, and who clinched the McDonalds account'. The deal was simple: the client organized a 'pitch', and the ad agency got an 'account'.

Typically, the agency would then receive an annual budget to handle that account, and tried its best within that budget to develop the most creative and effective advertising for its client. At the end of the year, the budget was depleted, and then replenished.

The most important feedback mechanism wasn't necessarily the effectiveness of the advertisements, as 'effectiveness' was difficult to quantify, but rather some crude alternative that measured the prestige (and thus, one supposed, the reach) of a particular agency or campaign. In establishing this norm, the ad industry was able to create a self-grading (and self-congratulatory) awards system; a system that graded ad campaigns on the basis of creativity instead of effectiveness and was arbitrated by other industry insiders instead of by customers or end-users. As a client within this system, it was great to have an industry award to put on your mantelpiece, and for the agency, awards were a must.

ENTER THE DIGITAL SCENE

In the last decade, digital has played an ever-increasing role in advertising. It is completely reshaping the business. Not only has the advertising industry finally found a mechanism to turn from mass-orientation towards individual one-to-one marketing, but also we can now measure the effectiveness of advertising down to three digits after the comma. Oops.

Let's examine those two game changers for the ad industry in a little more detail.

First, there now exists the ability to 'zoom in' on a particular profile, instead of having to use 'weapons of mass advertising' to target your customers. Over the last ten years, the Internet scene has evolved from simple websites that were nothing more than traditional brochures put on a new Internet-based platform, to interactive, engaging and personalized environments that present the right information, and the right functionalities, to the right profiles. The Internet is a mass medium, but allows companies to engage in almost tailor-made, hand-in-glove interactions and dialogues with their consumers. Great!

But ad agencies don't like it. The reason is that they perfected the art of the mass-medium to its pinnacle. They had polished the 'glossy full page ad in Elle magazine' as the apex of their craft. They were able to control, create, and dictate tastes, rather than react to existing preferences expressed by a wide range of individual customers. No longer.

Customers want interaction. Customers need personalization. Customers demand customization. Consumers now feel that advertising should be about them personally, and not the vague and overly-broad demographic category that advertisers had previously placed them in. The benefits of this new paradigm are huge: it is much more valuable for advertisers to reach one interested potential customer than it is to irritate ninety-nine others.

INSTANT FEEDBACK

I regularly give lectures to the ad industry. This industry is now in full, soul-searching mode. Advertising has gone digital and it is quite interesting to observe the almost frantic introspection to understand what the 'New New thing' is going to be. The ad industry doesn't like the concept of the New Normal, because for them it is fundamentally a game changer.

This is especially true when we combine the transition towards niche marketing with the second big blow to the industry: total transparency. In the last chapter we established that the digital world is increasingly able to capture, store, and recall data in real time.

Effectively, we can now measure *everything*, and thus at any given point in our interaction with a customer, we can create total visibility about the effects of a campaign. What a shock compared to the good old 'Mad Men' days.

The 'feedback cycle' in those days was extensive and delayed. You had a budget, designed a campaign, created an advertising approach, launched it, and then waited for the outcome of the sales figures of the products featured. These cycles took months, quarters, or even years before they revealed anything meaningful to the client. Sometimes the results were inconclusive and there was no judgment passed on the efficacy of the campaign (beyond, of course, the ad award system).

Today, the feedback is instant. You launch an ad campaign on a website and instantly see how many people are clicking on the banner, engaging in your dialogue and being captured by your content.

Total and complete visibility.

The ad agencies obviously don't like this development either. It has become much harder to hide bad work. Today as never before, clients are able to assess the handiwork of their ad agencies in real time, and many are shocked by how few customers or prospects have visited their million-dollar website, and how poor their customers' experience on that site has been.

JOINT ACCOUNTABILITY

Today I see plenty of corporations turning the tables on the advertising industry. Companies are demanding a shift away from 'budget' thinking, and are talking about joint accountability. About sharing the upsides and downsides, about 'partnering' instead of contracting. It means that ad agencies have to abandon the cushy, relaxed lifestyle they have enjoyed for a very long time, and the pressure to perform is now on them.

This connects with a consistent theme across the New Normal. A continuum of services is blurring the defined boundaries between companies. As CEOs become increasingly accountable to shareholders for quarterly numbers, they are passing this scrutiny down the line to service providers, who now work on a shorter leash and a tighter budget. One result is the confluence of staff across companies at the project level.

Another is the tendency toward win/win deals, because in a measurable, quantifiable world, consistently delivering win/win economics is the only way to sustain a business.

There are timing issues as well. I recently heard a Fortune 100 CEO say: "The problem is also speed. It used to be simple: we relied on the ad agencies to pick up trends. They would see a trend, pitch it to us, then we'd agree and they would go and do a campaign on that. But that doesn't work anymore.

In today's digital society, by the time they pick up a trend and bring it to us, they're too slow. So we have to invest ourselves, and rely less on the agencies."

I can imagine that the ad executives watch 'Mad Men' with nostalgia, and perhaps a tinge of regret. Life sure was easy back then. But in the New Normal, there is no hiding. Everything is digital, so everything is measurable. There is no murkiness in the New Normal. It's about total accountability.

 ABANDON ABSOLUTE CONTROL

The final rule I have for living in the New Normal is the notion that we will have to abandon the concept of 'absolute control'.

I grew up in a world where there was a fundamental belief in 'total control'. Many of the systems we have in place today – governments, corporations, and schools – are firmly rooted in this total control paradigm.

But the next generation looks at the issue of control from a different perspective. Generation Y has experienced a digital world driven by bottom-up thinking (Wikipedia), and network thinking (Facebook), and has very little rapport with the old top-down models. As we have seen in Chapter 2 in the limits of privacy, digital natives are also willing to give up control of personal data much more readily than digital immigrants.

The New Normal will make us think differently about control.

FROM THE BOTTOM TO THE TOP

Allow me to illustrate.

When I graduated from University in the 1980s, there was an economic depression and jobs were in high demand. I was happy to find a place at Alcatel, then a giant in the telecommunications infrastructure business, and a fine place to start as a young telecoms engineer. I'll always remember the first day at work when we were introduced to the company by a team from Human Resources. The most important thing they wanted to show us on the first day was the org-chart.

This is your boss. This is your boss' boss. This is your boss' boss' boss. And all the way up to the CEO. They showed me that from the top to the bottom in this company, there were 11 layers of bosses between the Chief Executive and me. All neatly laid out in an org-chart to show exactly what a tiny little shrimp I was in this massive corporation.

And when we were about to head home after a day of org-charts, I remember the head of HR saying to us: "If you work really hard, you can make a promotion every four years." Apparently he thought this was a motivational remark, but when I walked home that day I was thinking: "11 × 4 = 44 years. 44 years! I'll never make it." What a depressing way to start a new job.

Enter today's workforce. They don't care about hierarchy. They don't care about org-charts. They don't care about career planning. The next generation knows very well that where you are on an org-chart is essentially irrelevant, because in today's fluctuating economy, companies will shift, merge and acquire so often that an org-chart isn't worth the paper it's printed on. Companies have become dynamics, not statics.

The next generation knows that the key to their careers is the knowledge they accumulate and the experience they gain. That will define whether they are relevant to an organization, as opposed to their position in a box on an org-chart. The next generation knows that they will shape their own career, flowing from one 'project' to the next 'engagement'. Unlike the previous generation, a career path in the New Normal will not be controlled by a lifelong engagement with a handful of large corporations.

The world of the New Normal is a realm where bottom-up behavior thrives.

SELF-CORRECTING MECHANISM

Encyclopedias are great indicators of generations. My parents had a real encyclopedia. They paid an insane amount of money for the Encyclopaedia Britannica, and it was a trophy possession in their home, displayed with pride on the bookshelf in the living room for all visitors to see. Not touch. I recall the horror of my parents when we wanted to use the encyclopedia, because we 'might leave a dirty fingerprint' on that valuable asset. For my parents, it wasn't just about information. It was about status and value. They saw the encyclopedia as an investment. At one point in their lives, they had to make the difficult choice between a third child and the Britannica, and they had decided to go for the 27 books.

When I grew up, I bought a Microsoft Encarta CD-Rom. I wanted to have the accumulated knowledge of human history on a single disc. But Encarta was slow, very limited, and primarily absurd because the moment the CD-ROM got printed, it was outdated. The disc-based encyclopedias were just as 'dead' as the paper-based volumes my parents had bought.

My kids use Wikipedia. And Wikipedia has more information than any encyclopedia has ever had. There are over 15 million articles on Wikipedia, and there are articles written in 262 languages. More knowledge has been gathered here than in any other medium in human history, and the most amazing thing is that it grew bottom-up. There was no 'boss', there was no 'hierarchy'. It is basically built by the community. And the majority of the work of reviewing, challenging and shaping is done through peer reviews. That is typical of the New Normal.

Wikipedia is never 100% correct at any one moment in time. My favorite Wikipedia story is when the new pope was elected in 2005. The person previously known as 'Cardinal Ratzinger' from Germany was elected as the new pope, and when the white smoke came out of the Vatican chimney, he was denominated as 'pope Benedict the Sixteenth'.

The very moment he was elected, you could hit Wikipedia and type in 'pope Benedict 16' and you landed on a Wikipedia page that showed a picture of Darth Vader.

It was there for about an hour, until someone said, "No, the new pope doesn't look like Darth Vader, he looks like Cardinal Ratzinger", and changed the picture. This is the self-correcting mechanism at work, and overall it works pretty well.

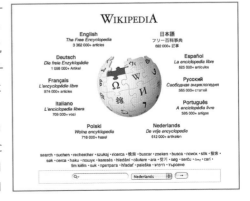

According to an investigation reported in the scientic journal 'Nature' in 2005, scientific articles in Wikipedia come close to the level of accuracy of Encyclopaedia Britannica and have a similar rate of 'serious errors'. And so Wikipedia is an example of good enough technology in the knowledge sphere, and the end of 'absolute control' thinking.

Shifting control from man to machine is something we have been doing steadily over the past decades and is a trend that will accelerate rapidly as we head into the New Normal.

I was flying from the US to Europe a few months ago in an Airbus, sitting next to one of the engineers from Airbus Corporation. He was an interesting seat mate to talk with, telling me that the plane we were in (like most Boeing aircrafts), has the capability to take-off and land with full automation in virtually any airport in the world, without *any* intervention from the pilot and has had this capability for the last ten years. Add the automatic pilot that flies a plane from A to B, include automatic take-off and landing, and you begin to wonder why you need the pilot at all.

Well, not really. Remember the amazing and extraordinary landing of US Airways Flight 1549 in the Hudson River in January 2009, when a brilliant pilot (Chesley 'Sully' Sullenberger, then 57, a former fighter pilot) was able to save 155 human lives because he had the capacity to act in an emergency? The Airbus computer is today incapable of reacting to an unpredicted event (in this case, a flock of Canadian geese that flew into the plane's engines).

But the 99.99% of all the other flights that are 'normal', could be completely handled by the computer in the plane. As a matter of fact, they could be handled better. The engineer from Airbus told me that the 'automatic' take-offs and landings are safer, more fuel-efficient, and more smooth than a manual landing. But the pilots refuse to let go of the stick, presumably because they are reluctant or even scared to give up control.

When I grew up as a kid, a plane flight was still an awe-inspiring experience. Somehow, we as human beings had coaxed hundreds of tons of inert metal to take flight. In today's world, the luster of air travel has tarnished, in part because of cost-cutting, but mainly because the wonder of aviation has been replaced by a banality that inevitably takes hold after an amazing thing is repeated over and over again. But in my childhood, when the plane had landed, and we had experienced a smooth landing, the whole plane would often burst out into applause to congratulate the pilot on their great touchdown.

I had an interesting experience on a recent trip from Newark to Brussels. The pilot was extremely talkative over the PA system and not only explained in detail how high she was flying, when she was taking a right over Greenland and how cold it was outside, but when we neared Brussels she informed the passengers that on that day "the plane would completely land itself, and all she would do is sit back and observe to oversee if all was going smoothly."

When the plane landed, it was magical. It was a *perfect* touchdown. No bumps. No noises. No shaking. It was as if the plane kissed the runway, gliding onto the tarmac of Brussels airport, a completely velvet transition from airborne to ground.

And it wasn't just me that had noticed it. The passengers on the plane burst out into applause that I instantly joined, perhaps a nostalgic childhood reflex.

And while I was clapping and applauding the great touchdown, I thought to myself: "Who am I clapping my hands for?" The pilot, who sat there and 'observed'? The plane for landing itself? The Airbus engineering team that had designed this great system? The programmer who had developed the code so the plane could land itself?

The New Normal is a place where we have to abandon the old 'absolute control' thinking. We will not be able to exercise absolute control over companies, consumers, employees or even experiences. Instead, we will have to embrace the idea that technology will allow us to work autonomously, independently and intelligently. Those are big steps to take.

CONCLUSION

Are these the only rules for the New Normal? Not by a long shot. There will be plenty more to deal with. You will have to decide what additional rules will govern your context, your market, and your organization as we transition to a new digital world.

I love the way that Douglas Adams reflected on the shaping force of technology, and how we react to it. In 'Salmon of Doubt', a collection of stories found on his computer when he passed away, he describes his rules for dealing with technology, depending on which group you belong to:

> Douglas Adams,
> *Salmon of Doubt*,
> 2002

0 → 15   Anything that is in the world when you're born is normal and ordinary and is just a natural part of the way the world works.

15 → 35   Anything that's invented between when you're fifteen and thirty-five is new and exciting and revolutionary and you can probably get a career in it.

35 → ...   Anything invented after you're thirty-five is against the natural order of things.

I don't know how old you are, but I'm quite sure we will *all* need to learn to deal with the New Normal and its new set of rules. My advice is to take the time to reflect on what those rules will be.

It was impossible to describe the motions of the planets before Newton had the sense to lay down the ground rules of the universe. First the rules, then the solutions. What will be the 'Laws of Newton' in the New Normal? Which ones will shape your future, or the future of your company? Take the time to think about the 'new rules' in the context of your organization.

**Then it's time to talk about defining a strategy,
or a set of strategies, for the New Normal.**

35 → ...
Anything invented after you're thirty-five
is against the natural order of things.

CHAPTER
04

CUSTOMER STRATEGIES FOR THE NEW NORMAL

"Because its purpose is to create a customer,
the business has two – and only two – functions:
marketing and innovation. Marketing and
innovation create value, all the rest are costs."
— Peter Drucker

"It will work. I am a marketing genius."
— Paris Hilton

"Marketing is too important to be left to
the marketing department."
— David Packard (co-founder of Hewlett-Packard)

If you look through old magazines, all of the wonderful pictures and photographs of both Time and Life are now fully searchable online. It is wonderful to observe the evolution of coverage on certain trends and topics.

> http://www.time.com/time/coversearch/

> Since November 2008, Google has hosted Life magazine's photographs. Many images in this archive were never published in the magazine. The archive is accessible through Google image search:
> http://images.google.com/hosted/life

Time, April 1993

Time, July 1994

Time had already been reporting on the digital revolution for a number of years, but the first Time cover that really hit a nerve was in **April 1993**: "The Info Highway: Bringing a revolution in entertainment, news and communication." Note that we didn't talk about the 'Internet' at that stage; it was still the 'info-bahn' or 'information superhighway' that would transport thousands of channels of content your way.

A little more than a year later, in **July 1994**, the word 'Internet' debuted on the Time cover "The strange new world of the Internet: Battles on the frontiers of cyberspace." In one year, the metaphor had shifted dramatically, from a 'highway of content' to a 'strange new world'.

The visual contrast was no less stunning: the 1993 cover was filled with the familiar elements of the human eye and a film-like strip of content, while the 1994 cover featured an other-worldly landscape and what today looks like flying, early prototype iPads.

This was when I was just starting to wake up to the Internet. I was already working in the telecom industry at that time, but still had friends who were in university when the World Wide Web tsunami hit academia. They got very excited about the possibilities of the Internet, and introduced me to the first browser and web pages. I was duly un-impressed. I vividly remember them getting feverish about the fact that they could go to a website that showed the weather in Chicago, and I remember thinking: "Why would I want to know the weather in Chicago?"

Despite my initial reticence, I latched on very quickly and soon I was the feverish young kid in the office trying to convince everyone to go online, while they wearily shook their heads thinking "He'll never last long." I didn't. I quit my job in 1995 to start my own Internet company, and never looked back.

In **March 1995**, the Internet exploded into general public consciousness. Time magazine's March 1st cover said: "Welcome to Cyberspace: Enter Here." But the Internet was still being positioned as something alien, and as a 'different place'. Around that time I bought the 'Atlas to the World Wide Web', a 200-page super-sized atlas in the Rand-McNally style of road maps I had known as a kid. The web atlas was an absurdity, outdated the very moment it was printed, but it was a wonderful illustration of trying to use old metaphors to describe something completely new.

OLD METAPHORS

Describing new technology with old metaphors is a common practice, because old metaphors provide a frame of reference for consumers. When cars were introduced, for example, they were introduced as 'horseless carriages', because that was the existing paradigm. In fact, if you look at the first generation of automobiles, they were actually designed like old carriages, sans horse. For a very brief period, they even installed 'mechanical horses' in front of the horseless carriages just to 'even out' the metaphor.

Fig. 1.—AN ENGLISH HORSELESS CARRIAGE OF 1827.

Steam carriage invented by Mr. Gurney. London Observer, December 9, 1827 www.catskillarchive.com

Consider a more recent example: when Netscape erupted onto the scene in 1995, they were truly pioneering new technology with the advent of the browser. But they fell into the 'horseless carriage trap' because although they had new technology, they used an old metaphor of the software industry to build their business model. Netscape tried to 'sell the browser' and make money from selling a piece of software. Eventually they failed, in part because they never shifted their corporate mentality from a software focus to an Internet focus.

Google is a great example of a 'post-horseless carriage' way of thinking, recognizing the Internet as a platform. Netscape stuck to the old metaphors, and vanished.

An important distinction should be made between how to conceptualize a business model and product, and how to market it to consumers. Using old metaphors and frameworks to describe new products is an effective method of bringing consumers from one paradigm to another. Building products and business models based on old metaphors, however, is dangerous. A business has to innovate and then market that innovation to customers.

As Henry Ford said famously: "If I had asked people what they wanted, they would have said faster horses."

Time, December 2006

It wasn't until **December 2006** that Time magazine figured out that the Internet wasn't about something new or an alien territory, but that it was all about *You*. In the famous 'mirror' cover of 2006, the consumer made it to "Person of The Year", because "You control the Information Age." The tagline for the cover read: "Welcome to your world." The Internet transformed from a 'strange new world' to 'your world' in roughly one decade.

And that's where we are. There is no *new* customer in the New Normal. It is *You*. But *You* behave differently in the New Normal as a customer than you did when you were still living in the analog world. Companies have to interact with You differently in the New Normal than they did when we weren't submerged in the digital pool.

So how can businesses build a new customer strategy in the New Normal, a new strategy to interact with people like you?

If Time magazine records the pulse of society, then why did it take 13 years for society, including consumers and entrepreneurs, to recognize the true implications of the Internet? A big factor is what happened in the technology sector and the markets in those intervening years.

On December 5, 1996, then-Federal Reserve Board Chairman, Alan Greenspan, made a now-famous remark during a speech at the American Enterprise Institute, saying: "How do we know when *irrational exuberance* has unduly escalated asset values, which then become subject to unexpected and prolonged contractions..."

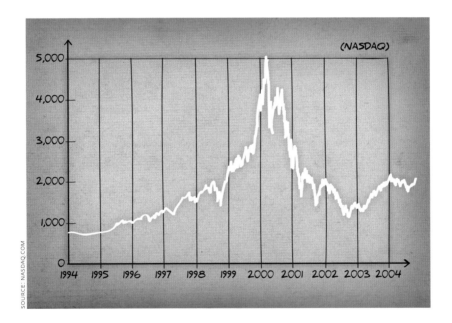

On the day after the 'irrational exuberance pronouncement', the NASDAQ Composite opened at 1275.90. Less than three and half years later, on March 10, 2000, NASDAQ peaked at 5132.52, a compound annual growth rate of more than 50%. During this period, popularly termed the 'dot-com bubble', thousands of companies raised billions of dollars from venture capitalists and the public markets, as investors and pundits dreamed of a 'new world' where the traditional rules of price-to-earnings ratios and profitability were supplanted by something else.

In retrospect, the Internet companies of the late 90s had, by and large, not created new business models, and had marketed effectively to investors

A BRIEF HISTORY OF BUBBLES

Bubbles are not that uncommon. The absolute reference on this topic is Charles MacKay's 'Extraordinary Popular Delusions & the Madness of Crowds'. Although the book was published in 1841, it remains readable today, largely because it describes human folly that has been repeated throughout the last four centuries.

Charles Mackay, *Extraordinary Popular Delusions & the Madness of Crowds*, with a foreword by Andrew Tobias, 1980

The first asset bubble on record is the Dutch **Tulip Mania** of the early 17th century. During this tulip-bulb hype, the price of a single bulb could exceed more than 10 times the annual income of a skilled craftsman. It sounds like complete lunacy to us now.

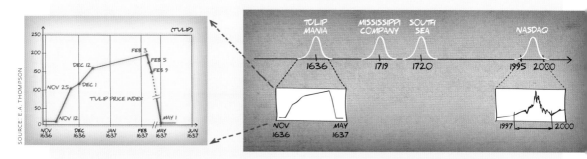

Unlike the Dutch Tulip bubble, which was characterized by speculation on physical assets, subsequent hypes were often financial asset bubbles very similar to the one we experienced in the first Internet hype.

The **Mississippi Company** bubble in France in 1719 was a rush to buy stock in the Mississippi Company, a corporation that had exclusive trading rights in what is now the American Midwest. Stock prices spiraled upward as investors clamored to get a piece of this exciting new territory, with seemingly unlimited resources and plentiful supplies. Speculators flocked to the Rue Quincampoix in Paris where the stocks were traded, and MacKay's accounts are enlivened by colorful, comedic anecdotes, such as the Parisian hunchback who supposedly profited by renting out his hump as a writing desk during the height of the mania.

Speculation was also the cause of the **South Sea bubble** in England. The South Sea Company obtained exclusive trading rights in Spanish South America in 1711. Immediately rumors about the potential trade in the New World were going around, causing a rush for the company's shares. The stock value of the South Sea Company rose instantly. When it appeared, afterwards, that the expectations couldn't be redeemed, the stock market crashed.

There have been a number of famous asset bubbles observed over the past century: the Florida Speculative Building bubble (1926), the Roaring Twenties Stock-Market bubble (1920s), the Japanese Asset Price bubble (1980s), the Asian Financial Crisis (1997), the United States Housing bubble (2000s), etc.

but not to consumers. The unexpected and prolonged contraction that Greenspan warned about ran from March 2000 till October 2002, destroying $5 trillion in market value and viciously blunting investor enthusiasm for Internet companies.

I was fortunate enough to have intensely savored the first Internet hype. The days of the first Internet revolution were absolutely astounding. Companies were rushing to bring their business to the web, users were flocking to every new sensation, and startups were running amuck trying to spend money quickly enough. Some people still have negative sentiments about the bubble, especially when it gloriously collapsed in 2000, but I think it was a rare privilege to have lived through a true bubble.

Looking back, history does seem to repeat itself, as investors get irrationally exuberant about a particular asset and bid its price up to unsustainable levels. Perhaps there is something at the nexus of human nature and the capitalist system that biases us towards such unbridled optimism despite a clear record of financial carnage. But for our purposes in charting a set of business strategies for the New Normal, the most interesting aspect of bubbles is their aftermath. What happens after a bubble bursts?

My favorite theory on this subject was developed by the Venezuelan thinker Carlota Perez in her excellent book, 'Technological Revolutions and Financial Capital: The Dynamics of Bubbles and Golden Ages.'

> Carlota Perez,
> *Technological Revolutions and Financial Capital: The Dynamics of Bubbles and Golden Ages, 2002*

Her theory holds that every 'major' technological revolution goes through five phases:

①	**Irruption**	decline of old industries and the appearance of new technology
②	**Frenzy**	sudden hype, intensive investments, and the exponential rise of the bubble
③	**Collapse**	complete breakdown of the bubble
④	**Golden Age**	when the real payoff of new technology occurs
⑤	**Maturity**	slow growth, and 'end' of the bubble period

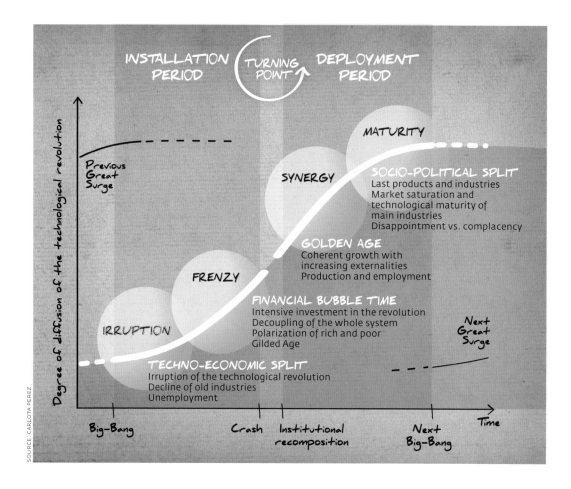

INSTALLATION PERIOD — TURNING POINT — DEPLOYMENT PERIOD

MATURITY

SYNERGY

SOCIO-POLITICAL SPLIT
Last products and industries
Market saturation and
technological maturity of
main industries
Disappointment vs. complacency

GOLDEN AGE
Coherent growth with
increasing externalities
Production and employment

FRENZY

FINANCIAL BUBBLE TIME
Intensive investment in the revolution
Decoupling of the whole system
Polarization of rich and poor
Gilded Age

IRRUPTION

Next Great Surge

Previous Great Surge

TECHNO-ECONOMIC SPLIT
Irruption of the technological revolution
Decline of old industries
Unemployment

Degree of diffusion of the technological revolution

Big-Bang · Crash · Institutional recomposition · Next Big-Bang · Time

SOURCE: CARLOTA PEREZ

Perez tests this framework by applying it to the Industrial Revolution, the railway bubble, the rise of electricity and the revolution of mass production.

The railways in Britain are a perfect example. When the railway concept was introduced, it sparked the 'Railway Mania' of the 1840s. This mania followed a very common pattern: as the price of railway shares increased, more and more money poured in from speculators, until the inevitable collapse of the asset bubble. It reached its zenith in 1846, when no fewer than 272 Acts of Parliament were passed, setting up new railway companies, and the proposed routes totaled 9,500 miles of new railway.

Around a third of the railways authorized were never built – the entitled company either collapsed due to poor financial planning, was bought out by a larger competitor before it could build its line, or turned out to be a fraudulent enterprise to channel investors' money into another business.

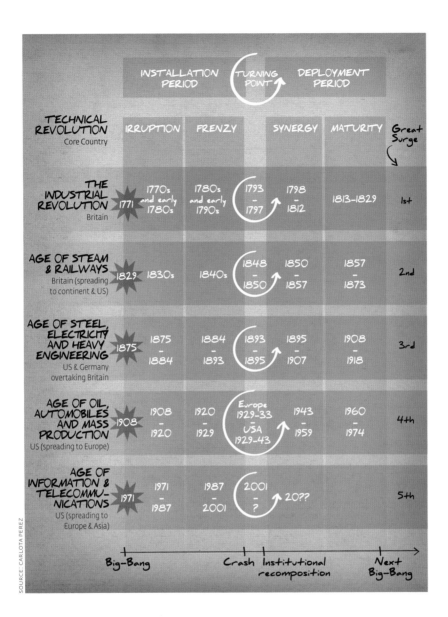

So what happened after the bubble burst? As predicted, it was in the aftermath that the 'golden age' of railways occurred, a period that was extremely profitable for those companies and businessmen that positioned themselves strategically as the railway system was built.

My reason for appreciating Carlota Perez is obvious:

The New Normal is set up to be the Golden Age of Information Technology, where 'digital' is normal, but the true impact of technology becomes apparent.

Companies have been extremely slow to understand the new dynamics of the Internet age. Back in the mid-nineties, the number one comment companies made when they decided they wanted a website was:

> "YES. Absolutely! We want one of those websites.
> Make sure it's one with lots of traffic."

We can chuckle at this sentiment because we know you can't 'command' traffic to your site (you can buy it from Google, but that's a different story). But is the understanding of most companies today more nuanced than it was 15 years ago? After all, in the 'social media frenzy' of Web 2.0, a typical comment from a company today is:

> "YES. We want to use social media. Make sure it creates a lot of traffic.
> And make sure they say a lot of nice things about us."

Looks like we're still adjusting to horseless carriages here. The New Normal isn't about *us* and *them* anymore when we think about consumers. It's about *You*.

GETTING INSIDE THE BRAIN OF THE DIGITAL CONSUMER

We have to make a big shift in the New Normal and stop thinking about digital as a form of new technology.

It's not about technology. It's about usage.

In Seth Godin's 'Purple Cow: Transform Your Business by Being Remarkable', we learn about technology and the creation of new markets through the story of Otto Rohwedder, who invented sliced bread in 1912. Rohwedder's first career was as a jeweler, and he became the owner of three jewelry stores in St. Joseph, Missouri. Leveraging insights gained from his mechanical work with watches and jewelry, he aspired to invent new contraptions, and eventually sold his jewelry stores to fund the development of a bread-slicing machine.

Seth Godin, *Purple Cow: Transform Your Business by Being Remarkable*, 2003

Rohwedder ultimately failed in his bread-slicing venture, not because he failed to make a good product (he developed a brilliant machine that produced beautifully sliced bread and even wrapped it), but because he failed to market it properly. It wasn't about the machine, and it wasn't about technology. The market simply wasn't ready for sliced bread.

As Godin concludes: "It wasn't until about twenty years later – when a new brand called 'Wonder Bread' started marketing sliced bread – that the invention caught on."

If we apply this to the New Normal, we have to get inside the brain of the digital consumer, and find out what makes it tick.

THE NEW GROUND RULES

For a long time, we've thought in terms of one key rule in the online space: 'content is king'.

When the Internet erupted, many companies believed that as long as they provided the right content, they would be *ok*. If businesses provided their customers or prospects with honest, decent and qualitative content, they would survive this revolution. This is why most companies essentially converted their existing paper brochures into digital format and called it a website.

Today, that long-standing rule is dead.

Content is no longer king.

If you consider the intake of information we process on a daily basis, and combine that with the limit of length which evolves to zero, the conclusion is simple: we can hardly rise above the noise. It's not that content isn't good, it's just that the digital consumer has been overwhelmed by content and is thus deadened to it, especially when you 'push' the content to the consumer.

The classic 'push' medium used to be television, and the classic 'push' paradigm was the mass market. In some ways, competition for consumer attention in the days of the mass market was like traditional warfare, with two armies lined up on the battlefield. Advertisers had a direct line of sight on a large group of consumers and honed their strategies based on this familiar medium. For example, in the 1960s, an advertiser could reach 80% of U.S. women with a spot aired simultaneously on CBS, NBC, and ABC.

The mass market has steadily eroded over the past four decades, and in July 2004, Business Week ran a cover story called "The Vanishing Mass Market", which declared: "New technology. Product proliferation. Fragmented media. Get ready: It's a whole new world." With the mass market fading, marketing is now akin to guerilla warfare. No more old rules. Anything goes, as long as the end goal is achieved; get close to the consumer.

How do you get close to today's consumer? Two telling quotes from the 2004 "Mass Market" Business Week article sum up the progression of consumer attitudes over time.

→ From the 1960s to the 1980s, advertisers steadily had to move away from the average consumer keeping up with the Joneses: "From the consumer point of view we've had a change from 'I want to be normal' to 'I want to be special'."

→ From the 1980s to the early 2000s, the mass market looked to increasingly customize: "Time, the prototypical mass magazine, long has been able to craft an almost limitless number of ad-customized versions of its national edition. We've done as many as 20,000 versions, but that's not something we want to do every week", says Publisher Edward R. McCarrick.

Is the consumer in the New Normal a new consumer? No. But he has a totally different behavior pattern vis-à-vis content. Content is no longer the main differentiator. Interaction is.

We are in the 'age of You'. The consumer is in charge. An excerpt from the 2004 BusinessWeek article predicted this well: "Companies must recognize that they increasingly have to engage gods and are not dealing with helpless consumers anymore", says Rishad Tobaccowala, an executive vice-president of Starcom MediaVest Group. "This is particularly true of young people."

If in 2004 companies were dealing with empowered consumers, in 2010 they are dealing with users who control their content destiny. We have actually gone full cycle, from the mass market push, to a digital pull – where you pull whatever content you want from the Internet by searching for it – to the individual push, where a user logs onto his Google reader or Facebook account, chooses the feeds/friends he wants, and then pre-screened content is pushed to him whenever new info is posted. From the old battlefield model to this new, exponentially more complex landscape, the field-manual on individual push marketing has yet to be written.

How, as a company, can you influence a consumer who constructs his own content world?

POTENTIAL CONTENT PRODUCERS

To start with, companies have to make their content available, existing in the same dynamic, digital universe as their target consumers. The current frenzy of interest in social media marks a concerted effort on behalf of business of all sizes. As BusinessWeek put it in 2009: "For companies, resistance to social media is futile. Millions of people are creating content on the social web. Your competitors are already there. Your customers have been there for a long time. If your business isn't putting itself out there, it ought to be."

However, creating a Twitter account is not like running a 30-second television ad for a number of reasons.

First, you may have lots of corporate digital content, but no digital followers, and you can't buy Twitter eyeballs like you could ABC eyeballs in the days of the mass market.

Second, you don't have static control over your content in the New Normal. A reTweet might flip a carefully calculated message 180 degrees.

Third, someone may be Tweeting about your company or product before you are. They may be a satisfied customer or a disgruntled employee, but in either case, they have a bullhorn just as loud as yours.

Traditional marketers have been flabbergasted by the social media phenomenon, because traditional marketers loved the old era of total control, and are terrified that thousands, tens of thousands, or even millions of people out there are not just playing the role of 'information receiver', but playing the role of 'creator,' or 'distributor' of content. And indeed, in the New Normal you now have hundreds of millions of potential content producers out there. Content alone will not help you rise above the noise. The worst thing to do as a marketer is to try to get back to your old comfort zone of absolute control. Control is dead.

'Learn and Leverage' is the new mantra.

In the 'age of you', consumer attention has become the most valuable resource. Advertisers need it. Innovative entrepreneurs are playing around with alternate business models, including directly paying consumers for their time. I love the wording associated with this new model of consumer interaction. An example comes from the company 'Youdata', which introduced a concept coined 'MeFile'.

www.youdata.com

Their slogan is great: "Advertisers are already buying your attention every day. They just aren't buying it from you. We can fix that." Their business model recognizes that your attention is a valuable product for advertisers, that you are the sole owner, and that you should therefore be the only one selling it.

The underlying economic reality is that as consumers, we trade our attention to content-producers in exchange for good content, and those content-producers in turn trade our interest to advertisers in exchange for money (presumably to fund the good content). The 'Youdata' model is not sufficiently robust to sustain a paradigm shift, but it could be an interesting gap model that succeeds in the transition from the mass market to the New Normal.

CONTACT IS KING. AND THE USER IS THE DICTATOR.

In the New Normal, it's all about *You*. Consumers will still want to be treated as kings, but will value the quality of the interaction as paramount. The new name of the game is 'contact'.

Survival in the New Normal is all about contact. Do not translate your current interaction pattern with customers to the web, but fundamentally rethink your interactions with customers knowing that they will become digital. Don't translate. Rethink. Just like shifting brochures to the web didn't work in the first Internet hype, shifting customer interactions to the web won't work either.

The reason why I'm reticent to use the term Web 2.0 is because most marketers have shortsightedly seen 2.0 as a wonderful set of 'solutions' to solve their broken interactions with consumers. But Web 2.0 isn't a set of solutions; it is a wakeup call. Most marketers have confused the symptoms with the cure.

LET ME ILLUSTRATE

I have been with the same bank for more than 15 years, however through merger, acquisition, and divestiture it is now a totally different entity than it was when I joined, with different staff, offices and branding. Over the years, my interactions with the bank have become increasingly digital. I rarely visit the bank branch and enact most of my transactions online. Despite these changes, which one might expect to weaken the connection between my service provider and me, I actually feel quite loyal to the bank.

I buy a wide range of insurance products through my bank, and also do my personal finance, business finance and my private banking with the bank, so I am a prime customer and in turn expect to be treated well. And I am, as long as I visit the branch.

A while ago I got a letter from my bank, informing me that they would raise our insurance premium because our house was too close to a river, which might cause a flooding problem. At the bottom of the letter was a phone number, an address, a fax number, and of course an email address with a request to contact them.

I sent an email, describing why the layout of our property would prevent damage in the event of a flood, and even included supporting pictures. Receiving no feedback, I thought "problem solved."

Three months later, I got the insurance premium in the mail; it had almost doubled. I called up my branch manager, who told me there must be a mistake, but then an hour later called back and explained: "I'm really sorry, but apparently we sent you a letter, and you never replied." They had lost my email. Could happen. But not in the New Normal.

The funny thing is that two weeks after this incident, I was invited to spend three days with the executive board members of the same bank, brainstorming "How can we be more Web 2.0 as a bank?" I couldn't resist telling my personal anecdote, which led to a refund in my account and, more importantly, a clearer understanding for the bankers about the relevance of a New Normal strategy.

DIGITAL LOYALTY

It is *not* about putting bells and whistles on your website, it is *not* about the facade and trying to facelift an old website with some fancy 2.0 elements. Here is where you have to separate the symptoms from the cure. The answer is to create *new digital loyalty* with your customers.

I'm still with the same bank. But I'm a digital immigrant. When digital

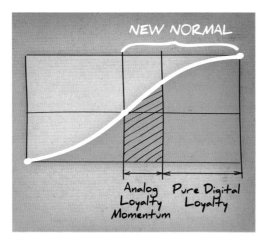

natives are the dominant consumer audience, they won't think twice about switching because they don't have any analog loyalty.

Because of the 'momentum' of analog loyalty, there is a period of transition in the New Normal where the old analog loyalty will still have effect, and the full impact of digital loyalty hasn't fully sunk in yet.

McLUHAN IN THE NEW NORMAL

I'm a big fan of Marshall McLuhan. He is the original thinker behind the concept of the 'global village', and his work is one of the cornerstones of our thinking about media today.

One of McLuhan's marquee expressions is: "The medium is the message", which means that the form of a medium embeds itself in the message, and that the medium itself influences how the message is perceived. The phrase, introduced in his 1964 classic 'Understanding Media: The Extensions of Man', suggests that we must consider the medium, as well as the content it carries.

Marshall McLuhan, *Understanding Media: The Extensions of Man*, 1964

In the New Normal, we have to revisit this phrase. The medium of the Internet has altered our behavior. In the New Normal, the medium isn't the message anymore.

In the New Normal, the response is the message.

The way a company will interact, the way a company will respond, the way a company will 'handle' the dialogue, *that* will become the determining factor in how companies are perceived and valued by the consumer.

THE EXPERIENCE ECONOMY

For a decade, we've been talking about the 'experience economy', first described by B. Joseph Pine II and James H. Gilmore in a 1999 book of the same name. Pine and Gilmore argued that the experience economy was the next economy, following the agrarian economy, the industrial economy and the service economy. If you look at the successful range of recent Apple products – iPod, iPhone, iPad – they are perfect examples of products that provide an intense experience for the end-user, and therefore generate an enormous amount of consumer loyalty.

> B. Joseph Pine II and James H. Gilmore, *The Experience Economy*, 1999

In the New Normal, where most of our customer interactions will be digital, every moment of every interaction is an opportunity to enhance the overall customer experience.

We mistakenly assume that digitizing interactions is a clear path to improved service, because interactions will be quicker and more convenient for the customer. This is only part of the puzzle. Although customers will expect the ability to interact with you at their convenience, on their terms, they will also expect the digital experience to be seamless and interesting.

And that's not simple. Making a digital interaction frictionless is hard work. As Dolly Parton once said: "It costs a lot of money to look this cheap." In the New Normal, there is a lot of work and effort required to make a digital interaction simple and enjoyable.

Some of the early winners of the Internet age know this very well. Take Amazon, a company that constantly tinkers with their award-winning website, often in subtle and nuanced ways, to constantly improve the user experience.

> www.amazon.com

Says Jeff Bezos, the founder and CEO of Amazon: "We see our customers as invited guests to a party, and we are the hosts. It's our job every day to make every important aspect of the customer experience a little bit better." eBay does the same with its auction website, as does Google with its suite of web-based products. Google actually promotes the concept of 'perpetual Beta' – a site is never finished, because it is constantly being tested, improved, and fine-tuned for optimal user experience.

This is why I shake my head when I see a press release from an airline or read a bulletin on some corporate website saying: "Welcome to our newly designed website." This is not going to cut it. In the New Normal, there are no more 'new and improved websites'. In the New Normal you are constantly optimizing the digital user experience, because the digital experience is your prime driver.

In the New Normal, beta is the new New.

New Limits!

LIMIT (CHANGES) = CONSTANT
LIMIT (FINISHED) = BETA
LIMIT (CUSTOMIZATION) = INDIVIDUAL

It is no longer about advertisers and media or average consumers. Forget 20,000 different editions of Time magazine. In the New Normal, the ratios aren't 1000 to 1 or even 100 to 1 but 1 to 1. It's about You. In the old days of advertising and marketing, 'they' called the shots. Now, you call the shots.

THE NEW PATTERNS

A passion for You is not enough. Recognizing that each individual consumer is the center of his own customized world is necessary to survive in the New Normal, but to thrive we also have to understand the new patterns that have emerged, and exploit these patterns to leverage our customers' digital experience.

The 'long tail' concept resonated immediately when Chris Anderson wrote about it in October 2004 in Wired magazine.

Chris Anderson, *The Long Tail: Why the Future of Business is Selling Less of More*, 2006

The concept is elegant and straightforward. If you look at a classic statistical distribution, there are two parts: the head and the tail.

For a long time, we followed a single strategy: focus on the head, and ignore the tail. Why? Because the tail was too difficult to mine, and the head was considered 'low hanging fruit'.

Anderson used Amazon as the quintessential example of a long tail company. The head is what every bookstore has in stock: the bestsellers that everyone wants to read (Harry

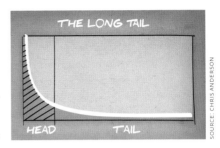

Potter, Da Vinci Code, Stephen King, etc.). Amazon, like every other bookshop, offered those titles, but Amazon also offered books that only a few people wanted to read, realizing that there were a huge number of niche titles of interest.

These are books on the World Exposition of Ghent in 1913, or silk handkerchief painting manuals, or… whatever. Although there are only a few people who want to buy these books at any point in time, they desperately want to buy them, and are willing to pay a premium.

Add all these people up and you get in the totality of the tail, which is actually quite significant. This is why Amazon was clever and successful; they took advantage of the head, as well as the tail.

Is the long tail a new concept? No. But the New Normal makes the long tail a concept that can be exploited profitably. For a long time, we used Pareto's Law – the 80/20 rule – and interpreted it as a focus on a limited set of customers who represent 80% of sales, and forget about the rest. In the New Normal, we know that this is no longer valid; although the head is low hanging fruit, the tail is where you make the difference. The tail is where you will find those customers who, if treated well, will turn into acolytes of your organization.

The tail is where you zoom in on the preferences, tastes and profiles of your most valuable customers.

The tail is the place in the New Normal where you can zoom in on You.

Amazon has been the poster boy of the long tail. But since Anderson wrote his article in 2004, virtually every market has been 'long tailed'. There are now examples of the long tail in the travel sector, finance sector, pharmaceutical sector, and more. Your imperative is to understand how to leverage the long tail for your organization, and how you can mine the underlying concept of zooming in on You.

PARTICIPATORS

If the old adage claimed that power corrupts, and that absolute power corrupts absolutely, then in the New Normal we could say that a small amount of power seems to motivate. That is, as businesses and institutions continue to lose power relative to the days of total control, individual consumers are increasingly empowered. This empowerment in turn breeds a culture of participation.

At each point along the curve of the market, from the head to the tail, we see a spectrum of participation.

In traditional economic theory, 'price discrimination' is a vehicle for extracting maximum value out of a marketplace by identifying consumers who are willing to pay more for a product, and then charging those consumers more. In the New Normal, 'participation discrimination' is a vehicle for extracting maximum productivity out of a customer base, by identifying consumers who are willing to contribute more to the community surrounding a product, and then providing them an outlet to do so.

An excellent example of 'participation discrimination' is the NikeID project. Nike allows you to customize your new sneakers and make a statement. The NikeID motto is: "You Design It. Nike Builds It." This neat program accomplishes two key goals: word-of-mouth advertising, and exposure to the long tail. Nike knows that there is an exceptionally loyal and influential group of 'sneaker freaks' out there that will fully embrace their newfound ability to personalize sneakers, and in return become even more loyal advocates for the Nike brand. These participators will also have a variety of niche interests, from regional soccer teams to hip-hop-inspired

www.nikeid.com

footwear, and so Nike achieves an authentic new line of niche shoes without having to hire new designers.

In the New Normal, we will have to accept the spectrum of participation, and accept that the user is not only more in control, but may contribute as well. Not everyone will design his or her own sneaker. Not everyone will write a Wikipedia article. But identifying, nurturing, and leveraging the participators can add a devoted army of designers and trendsetters to your team.

Take the example of the Lego group with their product 'Lego Mindstorms', a series of programmable robotics toys. Mindstorms were wonderful products on the boundary between traditional Lego bricks and Robotics, and quickly caught on with a very small set of nerdy Lego enthusiasts who saw this as a great opportunity to build all sorts of wonderful robots and contraptions with Lego pieces.

mindstorms.lego.com

These enthusiasts were not the typical 11-year olds tinkering with Lego bricks, but were often in their late 20s or older, working in the computer or telecoms industry, and still fond of their childhood Lego memories. They loved 'Lego Mindstorms'. They loved them so much that they took them to the next level. Before the company knew it, these super Lego Mindstorms enthusiasts hacked open the Lego Mindstorm system and improved and expanded on the original design. They were building stuff and using Mindstorm for things Lego had never dreamed of.

When Lego didn't have enough funding to develop a Mac version of the product for the K-12 market, it was one of the enthusiastic Mindstorm users who came to the rescue. Lego realized it had to leverage these super enthusiastic participators. So Lego endorsed the participation generation. Lego encouraged customer-extensions to the product line, giving hackers a license to extend its software and firmware and encouraging a healthy balance between company and consumer.

When Lego designed the next generation of the product, 'Mindstorms NXT', the company turned to its participating customers again. Lego recruited a small group of lead customers from the hacker community to consult with them on the design of the next generation product, and they followed the advice of the community.

Maybe you don't have those enthusiastic participators for your organization. Or maybe you just haven't found them yet.

In the New Normal, participators can no longer be ignored.

COMMUNITY

In the New Normal you will have to leverage the long tail in your industry, and you will have to adapt to participation behavior. But above all, you will have to realize that in the New Normal, the most important digital demographic shift is that your customers flock together and live in virtual communities.

Communities have been the lynchpin of human society since the beginning, with people of similar interests, ideas or beliefs grouping together. But the Internet has enabled us to dissociate communities from their geographical context, and people can now virtually gather in an online community and share common interests regardless of physical location.

You may be on LinkedIn or Facebook. Your children certainly are. Some of us have difficulty in adapting to online communities, and others thrive. Some people get hooked on online communities and become addicted to spending time in this virtual world.

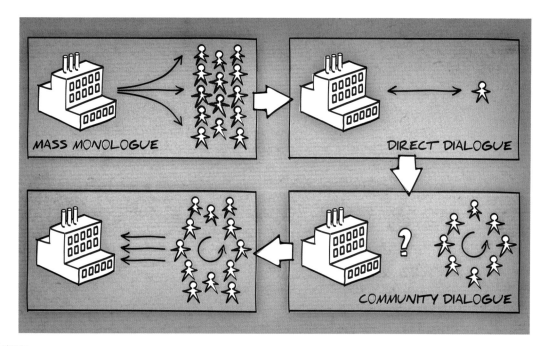

But regardless of your personal level of involvement in social networks, they are undeniably powerful. Today people are talking about products and brands online in communities, exchanging ideas on which product is the best and discussing their brand experiences.

If we consider the evolution of how companies interact with consumers, we've transitioned from mass marketing (step 1) to the Internet-enabled zoom in on *you*. This turns mass marketing into a direct (digital) dialogue (step 2), a form of two-way communication where the customer suddenly gains a bit of power.

The next step has been mindboggling to marketers: digital communities appeared and users started conversations amongst themselves. Customers started sharing experiences, and the Internet enabled the community to become a place for dialogue (step 3).

In the New Normal, the mass market is gone and has been replaced by communities. Some marketers may wish this new market would disappear, but in the New Normal, the way you talk to and involve communities will be one of your prime differentiators.

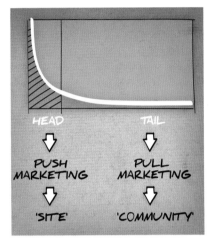

NEW BUSINESS MODELS

Combining the patterns outlined above will inevitably lead to new business models. The long tail is a new model, but some companies are going even further. One of the consequences of Chris Anderson's 'free' concept, which states that the limit of prices is $0, is the emergence of the freemium model.

Freemium is a combination of free + premium. If you plot this on the long tail, it means that the 'commodity' elements are being provided free of charge, but once you want the really nice features, the really great stuff, you have to pay. Just a little.

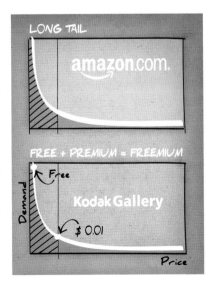

Kodak uses this model for its Kodak Galleries where you can share pictures with your friends, adapt pictures online and organize them into albums for free. However, as soon as you want to do more, like printing pictures or forwarding them to a wireless digital picture frame, you have to go via the web shop.

These aren't the only patterns. But the New Normal will fundamentally shift consumer behavior as well as engagement patterns between companies and consumers. In other words, the New Normal fundamentally reshapes marketing.

ORGANIZING FOR THE CUSTOMER IN THE NEW NORMAL

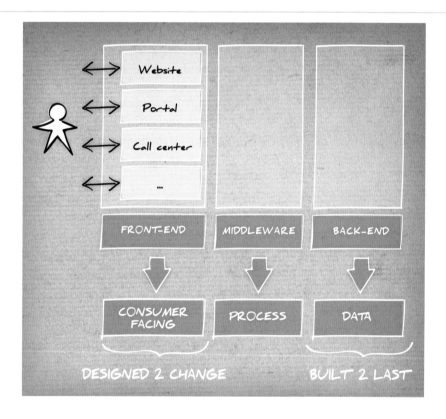

One of the 'old' concepts we still habitually use is the 'channel'. When the first Internet hype struck, we often wielded the term 'channel conflict', referencing a tension between the traditional channels (bricks and mortar stores for example) and the new 'direct' channels.

Today, most companies have a variety of different channels to communicate with their customers, and many of them are online/digital. Although we're moving into the New Normal, not all channels will be completely digital, as many companies will retain sales forces or call centers with human operators. But all companies will need to deal with multiple channels.

If we try and visualize this, a very simple, high-level view of the 'architecture' of how we organize customer interactions works well. A traditional way to view this is to look at the front-end, back-end and middleware components.

→ The **front-end** are the customer facing components. Here you find websites, portals, communities, but also traditional functionalities like call centers and sales force automation software.
→ In the **back-end** you find applications that drive the company's main processes, like billing systems, financial systems or ERP systems.
→ In the **middleware** you find the intelligence to tie the front-end and back-end together.

In the Old Normal, companies invested in back-office type applications, and each time there was a new channel (a new website, for example) organizations would bolt on this new channel in the front-end. Very often, companies developed quite a lot of intelligence in this new channel, but wouldn't want to spend much time on connecting the channel to other applications.

In the New Normal, we will have many channels, most of them digital. Many channels will have a temporary function, and will change rapidly. The rise of social media, for example, represents a whole set of new channels, with Facebook and Twitter suddenly becoming relevant to every business that wants to participate in customers' ongoing dialogue.

The problem is the way we built our channel intelligence. When we started building websites, we needed web content management systems to run this channel. When we started to communicate with email, we needed email relationship management systems to run this new channel. When we built call centers, we had call center software to run that channel. And every time, we started building 'intelligence' into the channel.

As a result, we knew our 'web customers' in the web channel, we knew our 'email customers' in the email channel, and we knew our 'call customers' in the call center channel. But this way of working no longer makes sense. In the New Normal, the customer is *you*. And *you* don't care if you're in a web database, a call center database or an email database. You want to be helped. You want to be served. Now. Social media is the new channel today, but what about tomorrow? We have to change our approach to managing customer interaction, and we have to change it quickly.

In the Old Normal, we probably invested too much in the channels. We built new channels the way we had built the old, back-office systems. Those systems were 'built to last'. And many of them are still lasting; ERP systems that were designed and developed in the 80s and 90s still run today. But new channels are different and have a much shorter shelf life. Many of the new channels are here-today, gone-tomorrow channels. Right now they are absolutely critical to your interaction with the customer, but soon they may no longer be.

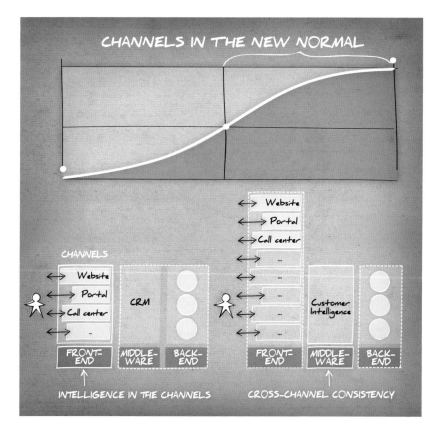

We have to rethink the way we design channels. In the New Normal, we have to envision channels that are 'designed to change' instead of 'built to last'. Websites are a good example. I can look at a picture of a car, and instantly say: "That's a seventies car." But the shelf life in the online world is much shorter. We can now look at a website and say: "Oh that is so 2004." The relevance of digital channels is huge, but the shelf life is much shorter.

The most important thing in the New Normal is to build the intelligence in your organization to interact with customers in a cross-channel, multi-channel reality. In the Old Normal, we would call this CRM: customer relationship management. Peppers and Rogers brilliantly articulated the whole field for customer relationship management in their book 'The One to One Future', which derived its mantra from the simple statement that it's cheaper to keep an old customer than it is to get a new one.

Don Peppers &
Martha Rogers,
*The One to One
Future*, 1996

END-TO-END INTEGRATION

In the New Normal, we must combine three different elements. We need end-to-end interactions with our customers. We need to work with cross-channel consistency. And we have to be able to zoom in to exploit the long tail.

We've covered the long tail already. For most companies it means rethinking the 'granularity' of customer interaction and zooming in on You.

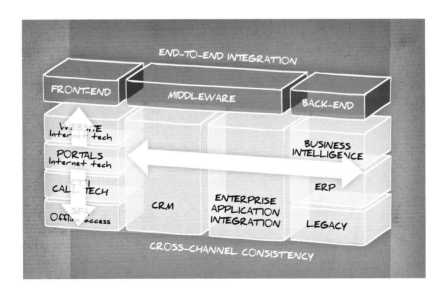

But the other two elements are equally important. The first is end-to-end interaction – as a New Normal customer, you don't care what channel you use, but you want the whole picture. In first generation websites, the front-end was often a 'facade' that wasn't connected to the back-end system. So if you filled in your details or placed an order, this information wasn't automatically connected to the back office systems to keep your records, and create and ship your order. This 'delay' between the front-end and the back-end was *ok* in the early days, but is frustrating to end users in the New Normal.

The expectation today is end-to-end connectivity.

The second is cross channel consistency. Customers in the New Normal will expect many channels for interaction, but they won't tolerate inconsistency in the information provided. A bank has multiple channels. I can go to its website (channel #1), or its portal (channel #2), call its call center (channel #3), or send an email (channel #4).

A digital native might even want to reach the bank via Facebook (channel #5) or Twitter (channel #6). But if I go to the website, fill in a pension plan simulator, download the results, add some details, send those back to the bank in an email and call 10 minutes later to discuss, do you think I will find seamlessly integrated channels? Probably not.

The customers of the New Normal require cross channel consistency, end-to-end interaction, and want the long tail to be tailored to their needs. Only by combining these three elements can we be relevant to our customers in the New Normal.

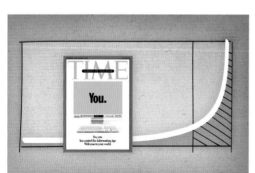

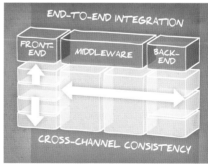

As we described earlier, the intelligence necessary to do this will have to operate in real time. You don't have time to process data overnight. When a customer uses one of your channels, you have to act. Instantly.

In the days of the first Internet hype, I remember the magical word was 'hits'. What we want to do today is to understand the customer rationale and delve into the customer intelligence. The companies that will be able to deliver real-time, cross-channel holistic insight and customer intelligence will prosper in the New Normal.

CONCLUSION

For most companies, the last ten years were about getting online. The next ten years will be about using the internet cleverly. Once we cross over into the New Normal, the way we interact with our customers will change completely. The worst mistake you could make is to take your existing analog interaction patterns with customers and transfer them to the web. Don't transfer, but re-think. It's all about intelligence. It's all about *you*.

As the master of marketing, Philip Kotler once said: "Marketing takes a day to learn. Unfortunately it takes a lifetime to master." You've got a couple years, max.

THE RIGHT QUESTIONS

① What is your digital loyalty strategy ?
② How will you leverage customer participation ?
③ What is your content-to-services strategy ?
④ What is your community dialogue strategy ?
④ What is your long tail mechanism ?

CHAPTER
05

INFORMATION STRATEGIES FOR THE NEW NORMAL

"As a general rule,
the most successful man in life is
the man who has the best information."
— Benjamin Disraeli

"One of the effects of living with
electric information is that we live habitually
in a state of information overload.
There's always more than you can cope with."
— Marshall McLuhan

"The fog of information can
drive out knowledge."
— Daniel J. Boorstin

THE VALUE OF INFORMATION

At the risk of overexploiting quotes, as Francis Bacon once said: "Knowledge is power." The right information at the right time is crucial.

That information is important is hardly new, but in the New Normal, the rise of digital information we have at our disposal becomes troublesome as we confront exponential runs of erupting content. If we don't change the way we look at information and learn how to deal with overload, the over-abundance of information could ruin us.

One of the great authentic thinkers in this field has been John Naisbitt. He was downright cantankerous when he wrote 'Megatrends' in 1982, and he predicted an avalanche of information – an avalanche so great and encompassing that we would be engulfed by it. As he put it: "We are drowning in Information, but starved for knowledge." Thirty years later, Naisbitt's prediction couldn't be more accurate.

John Naisbitt,
Megatrends.Ten New Directions Transforming Our Lives, 1982

We are drowning in information, but starved for knowledge

We battle through the information avalanche every day. Plowing through piles of emails, trying to cope with attached documents, presentations and whitepapers, and confronting an army of urls directing us to websites with interesting references, comments and summaries. And let's not forget the tweets we follow on Twitter, the messages on Facebook and the requests on MSN to engage in even more information exchanges. And once we've left the office, we might find ourselves multitasking sifting through our voicemail messages to hopefully empty that queue of content, and scanning the latest messages on our BlackBerry or iPhone.

We've come a long way in just twenty years.

Remember the old 'inbox trays' that most of us had on our desks? Remember when every morning the company mailman would drop by and pile letters, magazines and manila envelopes into that inbox, and our main task was to empty it by the evening. No email. No web. No cell phones. Seems like a completely different world.

In the last twenty years, our information society has changed completely. Our exposure to information has risen dramatically, and our intake behavior has transformed beyond recognition. More worrying, the pace of information overload seems to accelerate as we enter into an age of 'information insanity'.

INFORMATION STRATEGY

Today, most companies and employees are hardly harnessing the power of information, instead they have become slaves of the information overload. Instead of businesses using information to generate knowledge, they have succumbed to information bureaucracy. Instead of exploiting the value of information, companies have become the victims of information.

The strange truth is that the information technology revolution that promised to help us get a handle on information actually put a clamp on our productivity.

And the promises were so spectacular. Remember the phrase "Information at your fingertips?" One of the companies that was among the earliest advocates of the vast potential of information technology was Apple Computer. Once they made their mark on the world with the enormous success

of the Apple II home computer, they set their sights on the corporate universe with the launch of the Apple Lisa in 1983 and the Apple Macintosh in 1984.

It was revolutionary. For the first time, the power of Information Technology wouldn't be in the hands of the IT elite, but would instead be at the disposal of the common people. Steve Jobs labeled the Mac 'Wheels for the Mind'.

Thinking back to when those early Macs and PCs first came out, it's hard to believe that they weren't connected to anything. No networks, no Internet, no Wi-Fi. Nothing. I for one remember spending my childhood toiling away for hours and hours on those personal computers, and yet I sometimes wonder what I actually *did* on them.

While the dawn of personal computing in the 80s started our journey towards information technology, the rise of the World Wide Web in the mid-90s turned the information revolution into hyperspace.

Our information hunger was fueled by increasingly available content and increased connectivity.

More than 25 years later, we can conclude that the advent of personal computing has completely changed our lives. It transformed the way we work, live and communicate. It's hard to imagine a world without Google, Skype, email or a web browser.

We have become addicted to information.

And yet, most companies still do not have an information strategy. They have information systems. Plenty of them. They have content management systems, intranets, document management systems, file-servers and email systems.

But most do not have an information strategy.

They have massive terabytes and petabytes of storage capacity. They back up millions of emails every night, and have huge archives of digital content, and yet staff can hardly locate anything. They have email systems that allow people to pump around enormous amounts of information from mailbox to mailbox, but users lack real mechanisms to retrieve the intrinsic knowledge of the organization.

WHO'S RESPONSIBLE?

We have survived, until now, without proper information strategies. But that time has come to an end. Naisbitt was right in 1982. Companies are choking on an overload of information and are starved for real knowledge.

Today we're at a turning point. In the New Normal, we have to get our heads around an information strategy.

The necessity for such a strategy is clear, but it's still hazy who should take this challenge forward. Until now, the answer has fallen between the cracks of management and IT. Management feels that since the challenge is about information, it should be the responsibility of the information technology people.

The IT people think that their primary role is to supply information systems such as email and document management, which allow the business to work with information. IT assumes management is responsible for ensuring that the information itself is handled and used in the proper way.

The result therefore is that in most companies, no one oversees the information strategy.

Information Strategy resides in some sort of No Man's land.

Furthermore, in the New Normal, the situation grows ever more complex. With the rise of consumerism, most employees have access to more advanced information technology outside the work environment than they do inside their companies. They have better equipment at home than at work. They have better Internet access at home than at work. And most use more advanced applications at home than at work.

As a result, their information behavior at home is vastly different than it is at work.

**In the New Normal we will have to deal with the fact
that our information behavior adapts faster
than we can implement information systems.**

DAY BECOMES NIGHT. NIGHT BECOMES DAY.

This has been a wakeup call for IT departments. For years, they introduced the shiny new toys and fancy applications, but that all changed with the advent of the World Wide Web. Most people had their first taste of the web at home. Using a simple modem, and with what now seems like antiquated material, we managed to navigate to our first webpages, and explore the first websites.

Then things started moving quickly. Instant messaging arrived and people started chatting online. Social media arrived, and people caught on to blogs and wikis. The rise of Facebook and Twitter gave a completely new dimension to interpersonal communication.

The IT department stood and watched and didn't have a clue what was going on. All of a sudden, this group of sullen users who had always rejected changes, and who had always opposed new technology, were suddenly bringing the wonderful technology that they had discovered at home with them to work.

They didn't need the IT department any longer.

Nonetheless, the IT department continued to work in a disciplined matter, with releases, schedules and planning. They worked in projects with proper project management. In classic projects such as the introduction of an 'Enterprise Resource Planning tool' or a 'Financial Accounting Package', the IT department dictated the rate at which the projects were executed. They controlled the flow of functionalities.

In the New Normal, however, it's the end user who decides what applications and systems they use. They decide which Internet tools to use, which applications to download on their iPads and iPhones, and which collaboration tools to use from the 'cloud', which has become synonymous with the myriad of functionalities available via the Internet.

Imagine the IT department's horror.

In a digital society, information behavior changes faster than information systems. And often IT can't keep up with the frantic pace.

FROM PUZZLES TO MYSTERIES

The Old Normal was all about solving puzzles. The New Normal is about solving mysteries.

In 'What the Dog Saw', a collection of essays by Malcolm Gladwell, there is one particular story that is incredibly relevant to the world of information and information overload. In the essay 'Enron, intelligence and the perils of too much information', Gladwell describes the difference between a puzzle and a mystery.

> Malcolm Gladwell,
> *What The Dog Saw*,
> 2009

A puzzle is a problem or a question with a definitive answer. Its solution depends on finding all the relevant pieces of information. Gladwell describes the Cuban missile crisis as a perfect example of a puzzle. All of a sudden, a picture surfaces of nuclear warheads aimed at the United States of America, and the Kennedy administration faces a serious crisis. However, the action to be undertaken is clear; get the missiles out of Cuba. It's a puzzle, because by gathering more information (intelligence in this case), the U.S. was able to find out exactly *how* to solve it.

A mystery on the other hand is an overload of information, and it's not easy to understand how to connect A to B, or even where A and B are. Gladwell refers to 9/11 as a perfect example of a tragic mystery. All the information about the bombing of the World Trade Center and the Pentagon were known to US intelligence officials. The identity of several terrorists was known, but the information 'fog' was so great, it wasn't clear there was a real threat. In essence, a mystery can't be solved with more information. Instead a mystery can only be solved by providing insight into the information, filtering the information and being clever with the information.

You might wonder why the title of the essay mentions Enron. As Gladwell explains, with the Enron debacle, the problem wasn't a matter of insufficient information; nothing was withheld. The information was all there (i.e. that the company was bankrupt) hidden within the mountains of financial disclosure.

The point Gladwell makes is that when faced with a problem, it's necessary to determine exactly what kind of a problem it is. If it's a puzzle, more research is needed. If it's a mystery, the solution calls for better analysis. Shrewd observers are needed to plow through existing material, re-analyze it, ask different questions, turn it upside down and expose the pattern that solves the mystery. Enron was a mystery, not a puzzle.

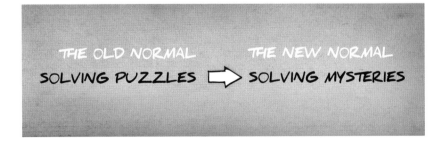

Until today, we've been puzzle solvers. Anytime there was a problem, an issue, a challenge in our organizations, we provided solutions by building information systems. That's why we have companies filled with information 'pockets', ranging from document management systems and network drives to intranets brimming with information. Plus there's the amazing collection of documents stored on our personal devices, hard drives, backup drives, phones and PDAs.

We have created spot-solutions for every information challenge we've encountered. We have created 'fishbowls' of content all over our companies, and there's very little interchange between the different information silos.

We constructed these information traps because we were solving puzzles.

In the New Normal, our challenge is to become mystery solvers. The mysteries of the New Normal information age have become quite critical. Mysteries require judgments and the assessment of uncertainty.

The difficulty is not that we have too little information but that we have too much.

A 'bit' is the smallest information unit for digital data. It's either a 0 or a 1.

A byte is 8 bits. And a million bytes are a megabyte. A thousand megabytes is a gigabyte. That's about enough to store one hour of normal quality video material digitally. For high definition video, a gigabyte would only get you seven minutes of video.

A million gigabytes is a petabyte. Last year, the entire world produced 800,000 petabytes of information. This year we need a new word, since for the first time we will produce more than a million petabytes. And that's a zetabyte.

A PETABYTE IS A LOT OF DATA

1 PETABYTE	20 MILLION FOUR-DRAWER FILING CABINETS FILLED WITH TEXT
1 PETABYTE	13.3 YEARS OF HD-TV VIDEO
1.5 PETABYTES	SIZE OF THE 10 BILLION PHOTOS ON FACEBOOK
20 PETABYTES	THE AMOUNT OF DATA PROCESSED BY GOOGLE PER DAY
20 PETABYTES	TOTAL HARD DRIVE SPACE MANUFACTURED IN 1995
50 PETABYTES	THE ENTIRE WRITTEN WORKS OF MANKIND, FROM THE BEGINNING OF RECORDED HISTORY, IN ALL LANGUAGES

As E.O. Wilson, an American biologist, said: "The world henceforth will be run by synthesizers, people able to put together the right information at the right time, think critically about it, and make important choices wisely."

E. O. Wilson, *Consilience: The Unity of Knowledge,* 1998

THE BASICS

In order to start exploring how to build an information strategy, let's start with the basics. The fundamental pillars of information are quite simple: Content, Collaboration, Intelligence and Knowledge.

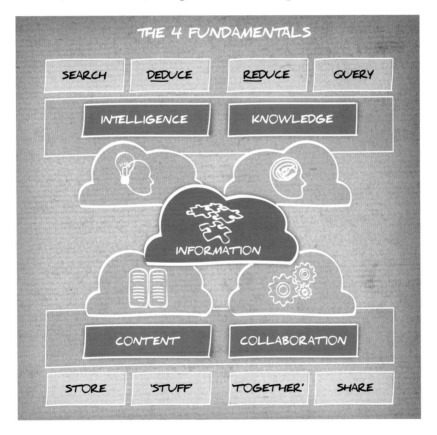

CONTENT

The simplest to understand is content. Content is just 'stuff'. This could be documents, presentations, PDF files, whitepapers, memos... anything. Anything someone would like to store, because essentially, what you do with 'stuff' is store it.

Where content is stored depends on your level of IT sophistication. This could be as straightforward as a folder on a hard disk, or could be as refined as a corporate document management system where documents have to be 'checked-in' with relevant information such as author, title, version number, etc.

Content isn't all that exciting. As a matter of fact, the science of content management hasn't evolved much in the last thirty years. Content is really only about producing a document and finding a place to store it so it can be retrieved later.

COLLABORATION

The second root element of information is collaboration. This is all about sharing. Collaboration means working with information, and increasing its value by sharing it with others. It's a way to work together, share and add value through the merits of cooperation.

Collaboration isn't a new field either. Older platforms such as Lotus Notes provided the embryos for collaboration in the early 90s. The advent of the Internet however has accelerated the pace of collaboration and allowed anyone the possibility to engage in instant collaboration.

INTELLIGENCE

Intelligence is the higher goal of information. What do we really want to achieve with information? Ideally, what we want is the collective intelligence inside these content vaults – these mountains of data and heaps of documents – exposed and available for all to benefit from it. The verb to use here is 'deduce'. We aim to 'deduce' gems of intelligence from an avalanche of corporate information.

The trivial, unembellished version of this concept is that we hope to have a 'Google' inside our companies. Although it is easy to argue that Google certainly is *not* the most intelligent way to access information, most are familiar with the concept of 'Googling' something to find out more. What we want is to deduce intelligence out of masses of information, and then allow users to search that intelligence.

de·duce [dɪˈdjuːs /ⓤⓢ dɪˈduːs]

arrive at (a fact or a conclusion) by reasoning; draw as a logical conclusion: *little can be safely* **deduced from** *these figures* | [with clause] *they* **deduced that** *the fish died because of water pollution.*

re·duce [rɪˈdjuːs /ⓤⓢ rɪˈduːs]

1. make smaller or less in amount, degree, or size, e.g. boil (a sauce or other liquid) in cooking so that it becomes thicker and more concentrated.

2. reduce something to change a substance to a different or more basic form: *it is difficult to understand how lava could have been* **reduced** *to dust.*

3. present a problem or subject in (a simplified form): *he* **reduces** *unimaginable statistics to manageable proportions.*

—— **KNOWLEDGE**

The final, and probably most noble of the four pillars is knowledge. This is perhaps the most underdeveloped area and the most overhyped for the last twenty years. Remember the concept of 'knowledge management systems'? Remember the promise that we would soon be able to query all the available knowledge inside our companies at the 'touch of a button'? Well, compare that to the documents stored in file servers with the unnavigable folder structures which we continue to back up every night on terabytes of storage.

Knowledge is about 'reducing'. A fine chef reduces a sauce by boiling the liquid to increase its density, concentrating it until it becomes a proper rich sauce. That's what we aim to do with information; condense it to its very essence in order to tap the rich knowledge inside.

When we want to search intelligence, we want to query knowledge.

PARKINSON'S LAW OF INFORMATION, AND WHY YOU NEED TO THINK IN TERMS OF QUANTUM MECHANICS

One of the mythical laws governing the world of information is called Parkinson's Law. This law essentially means that whatever information capacity you supply to humans, they will use it up.

> Parkinson's Law:
> Data expands to fill the space available for storage.

As we now know, the world of digital technology is governed by Moore's Law, which states that capacity doubles about every 2 years. We've had Moore's Law for over 35 years, and it hasn't slowed down. The capacity of computer chips, storage systems, memory, etc., doubles about every 2 years. And it looks like Moore's Law is here to stay.

> Moore's Law:
> Capacity in IT doubles about every two years.

But Parkinson's Law says that doubling capacity simply isn't enough. The best illustration of Parkinson's Law is a teenager. If you give a 14-year-old a brand new computer in the beginning of summer break, with the largest hard disk you can find, by September the hard disk will be completely full. That's Parkinson's Law. Whatever you give to humans, they will use. Of course, this not only applies to computer hard disks. Give people a drawer, they will fill it. Give them a closet, they will stuff it. Give people a garage, the next time you look there is no more room to park a car.

However, the results are quite catastrophic at a corporate level: if we could see what the IT department backs up every night, we would be astounded. The amount of information that is backed up only increases, never decreases. And the question is, which part of what is backed up every night is truly valuable or even necessary? Today there is a huge gulf between reality and perception on how we deal with 'old' information.

WE NEVER THROW ANYTHING AWAY

Probably the most underutilized feature in modern day computing is the trash can. It first made its appearance on the Macintosh in 1984. In those days, it was actually fun to drag something in and delete a file. They should have made it a lot more amusing; perhaps we would have used it more.

It's amazing how much information we produce. It's even more amazing how little we throw away. But then again, this is human behavior. People who have overstuffed closets of clothes may eventually start tossing out pieces they haven't worn in a year. That's usually a pretty good measure: if it hasn't been worn in a year, it probably won't be worn again.

The most underutilized feature of modern day computing.

In information technology, our closets are simply too big. We keep adding more and more capacity to our information closets, and back them all up. We never get to the point where we really have to throw out any old stuff.

I reckon, if you haven't looked at a document in over a year, that's usually a pretty good indication you never will look at that document ever again.

TOWARDS QUANTUM

We have come to a point in information strategy where there are two fundamentally different ways of looking at the world and – and this is a bit of a stretch – it's almost comparable to Newtonian and Quantum Mechanical thinking in physics.

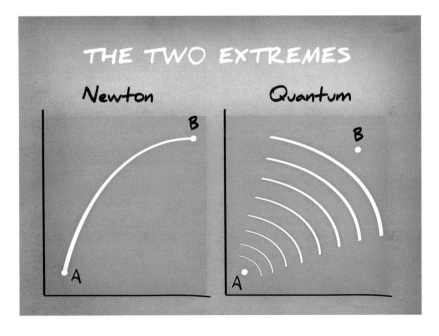

THE TWO EXTREMES

Newton   Quantum

Here's a little background:

Newton was an absolute genius. He was a believer in an absolute truth; an absolute possibility to understand the laws of physics and use that as the absolute guiding mechanism to explain everything in the universe. Remember the laws of physics that Newton wrote down in his 'Principia Mathematica': clean, simple and elegant. They allowed for the exact calculation of how planets revolved around the solar systems and how an apple falls from a tree. The choice word here is 'exact'.

> Isaac Newton, *Philosophiæ Naturalis Principia Mathematica*, Latin for "Mathematical Principles of Natural Philosophy", often called the *Principia Mathematica*, first published July 5, 1687.

But in the early 1900s, a group of physicists started observing not just the large objects, such as planets, that performed Newton's laws like celestial clockwork but also small objects like the inner workings of atoms, made up of tiny particles such as protons, neutrons and, even smaller, electrons.

Strangely enough, they couldn't explain those observations with the simple and elegant laws of Newton any longer. This gave rise to the field of Quantum Mechanics, which not only introduced science to a totally different and exciting new reality, but also sparked a mammoth controversy in the physics community.

Crucial and famous was the conference held in October 1927, the 'Fifth Solvay International Conference on Electrons and Photons', where the world's most notable physicists met in Brussels to discuss the newly formulated quantum theory. The leading figures were Albert Einstein and Niels Bohr.

With Quantum Mechanics it was no longer possible to know *exactly* where a tiny particle like the electron was, but instead there was a 'probability' field that described 'pretty much' where the particle would likely materialize.

Instead of an exact science, with exact laws, physics suddenly turned into a science of probabilities and statistics. Brilliant minds such as Einstein refused to believe in Quantum Mechanics at first. His famous statement "God does not play dice", was an expression of his disbelief that we would abandon the precise and exact laws of Newton.

The full story is that when Einstein, disenchanted with Heisenberg's 'Uncertainty Principle', remarked "God does not play dice", Bohr replied, "Einstein, stop telling God what to do."

In the course of the 20th century, it became clear that Quantum Mechanics was very real indeed. Once we start observing the very small, we had to factor in the uncertainties and probabilities to grasp the world of the subatomic.

The biggest jump here is we are now thinking in terms of 'probability' instead of 'certainty'. We went from a staid to a much more dynamic world. The choice word here is 'dynamic'.

QUANTUM THINKING IN INFORMATION

We have long believed it was possible to have absolute control over information, and that we had to build systems that gave us exactly the right information. But in the New Normal, we might have to surrender this belief and see if a more dynamic, but less controlled, way of looking at information is more appropriate.

Let me explain. 'Angels and Demons' was a cinematic adaptation of a bestseller by Dan Brown about a conspiracy in the Vatican. The main character of the film, played by Tom Hanks, is a Harvard professor who makes an enormous effort to gain access to the Vatican Archives. When he finally enters the Vatican Archives, they are brilliant preservation vaults that store the original copies of manuscripts from the last 20 centuries. They are a perfect 'Newtonian' abstraction of an information strategy: the main aim of these archives is to 'not lose anything'.

This is exactly what we've been doing in information strategy for a long time: making sure we back everything up, storing everything and above all, not losing a single document. Who cares if the users can't find anything, as long as we don't lose anything.

In the New Normal, we still want accurate information, but velocity is becoming more important. Accessing, searching and retrieving information quickly is more important than 'not losing anything'.

So, instead of building information systems like the Vatican Archives, we have to think differently. We have to focus on the rapidity and accessibility of information, the speed at which information is pumped around the company, and the swiftness of being able to query information becoming the dominating factors.

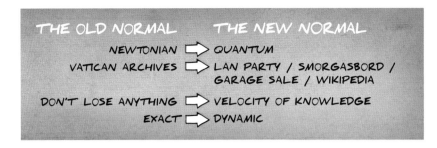

THE OLD NORMAL		THE NEW NORMAL
NEWTONIAN	⇨	QUANTUM
VATICAN ARCHIVES	⇨	LAN PARTY / SMORGASBORD / GARAGE SALE / WIKIPEDIA
DON'T LOSE ANYTHING	⇨	VELOCITY OF KNOWLEDGE
EXACT	⇨	DYNAMIC

A lot of companies today still have 'old school' information strategies. But what is the use of backing up 17 petabytes every night if none of the users ever look at it again?

The quantum view of information on the other hand is that the possibility of losing something no longer matters. It is more important that people find information quickly. In quantum thinking, swiftness is more important than actual perfection. Exact is nice. Dynamic is even nicer.

Of course you will still have plenty of Einsteins in your company who will refuse to embrace the new dynamics of information, and who will refuse to adopt a new mindset. They will continue to lament about every lost document and will tell you at every opportunity that you shouldn't 'play with dice'.

But the opposing concepts of exact and dynamic could possibly be combined. Perhaps you don't need to make an agonizing choice between the two. It all depends on your information strategy.

DRIVERS OF INFORMATION STRATEGY

Let's look at the six main drivers of information strategies.

The first three are the 'defensive' drivers of information strategy.

❶ COMPLIANCE

Many companies have to maintain records. They have to focus on storing information. They also have to manage information. There could be a number of rules, regulations and policies that force companies to have 'compliance' be a driver of their information strategy.

Some industries, such as the pharmaceutical industry, have strict rules that apply to information, since the submission of a new drug to the Food and Drug Administration has to follow stringent rules in terms of who can access, approve or change information. Some companies submit to rules such as Sarbanes Oxley (created after Enron) to insure they have rigid policies on the upkeep of information.

❷ CONTROL

Some organizations have prioritized the constant improvement of the quality of their information. They focus on improving the processes that govern information flow and want to follow the lifecycle of a document closely.

The lifecycle of a document encompasses the different stages that the document or other piece of information goes through, from creation to destruction. Different stages in the lifecycle can be Create, Approve, Publish, Archive, Destroy etc.

❸ ARCHIVE

If the main focus of information within the company is to 'not lose anything', then archiving is the driver. Like with the Vatican Archives, the company goes to great efforts to store information for long periods of time – not just documents but also data or records – and works diligently on mechanisms to be able to retrieve archived material quickly and easily.

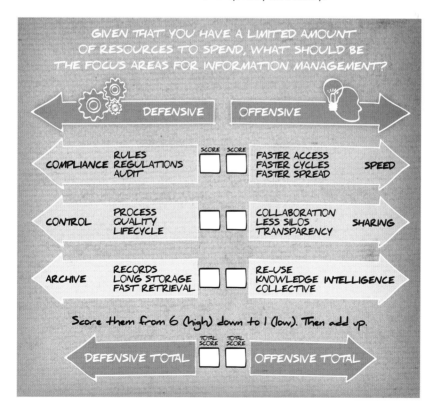

The next three are the 'offensive' drivers of information strategy.

❹ SPEED

The company wants faster access to information. It wants information to be 'pumped' around the organization at an ever increasing pace. It wants the cycles of information to get shorter and shorter, and the spread of information to increase.

❺ SHARING

The company wants to allow an environment where it's not only easy to share, but also easy to work together. It wants people to be able to co-create to produce more value in information. It wants to break down the silos of information and enable more transparency of information.

❻ INTELLIGENCE

Intelligence is about being able to re-use information. An environment is created where information can be 'built' upon, and intelligence can be 'constructed' going forward. The collective creation of knowledge is promoted within the organization.

Are these the only drivers? Probably not, but they are good for initiating a discussion inside the company. They are also a good starting point to evaluate where the company is now and determine what the organization should evolve into.

Some people might say: "I want all of these", but that would be impossible to achieve. A simple way to use these drivers is to think about what compromises would have to be made. Suppose you have a limited amount of resources (money, people, time) to spend, and yet you still need to design a robust information strategy. The question becomes: "Which driver do you focus on as a priority?"

The simplest way to determine this is to fill in the six boxes in the graph on the previous page, assigning 6 to your highest priority, 5 to the next one, and so on until you give a score of 1 to your lowest priority. Add up the left three boxes, and the three boxes on the right, and if your left total is higher than the right, you have a more 'defensive' information strategy. If your right-hand score is higher than the score on the left, you have a more 'offensive' information strategy.

You are your inbox. I believe that our inboxes, and how we maintain them, how we organize them, and how we store things in folders, say a great deal about our personal information behaviors and our personal 'information attitudes'.

There are those with thousands of emails in their inboxes, hundreds of them unread, and they are OK with that. On the other hand, there are people who have a compulsive drive to clean up their inbox to the point where they don't need to scroll anymore, and have an obsession to 'tidy up' their inbox before they can leave the office. There are those people who let their inbox flow over to a couple of hundred, and then suddenly get into a 'spring cleaning' frenzy, eradicating every single email until they have a pristine inbox; I'm one of those.

Companies also have an 'information attitude', and we have to learn how to understand it and perhaps see how we can eventually change or adapt it.

For a long time, I've been collecting the 'folder names' of companies I've visited over the years. What started out as a joke became a quite serious hobby, and now it has morphed into somewhat of an obsession. Call it a 'Folder Fetish'.

When I visit a company, I love to snoop around the 'shared drives' or 'network drives' (you know, the N: drive, the Z: drive, or the W: drive etc.) and see how creative people have been in naming folders. And they say a lot about their companies. A few examples:

I once found a folder on a shared drive that said:

PleaseDontDeletePleasePlease

This was obviously a company pretty strong on long-term archiving.
In another company I found a folder that read:

TempFolder_version7

Clearly an example of a company dead set on version control.
One of my favorites was a folder I found hanging around on a shared drive that read:

WeFoundThisStuffOnFredsHardDriveWhenHeLeft

Clearly an example of brilliant knowledge management at work here.

In a multinational company, I found:

CrapFromHeadquartersThatWeHaveToKeep

And my absolute favorite, and I won't mention the company:

DeleteThisFolderWhenTheAuditorsComeIn

I'm sure you too will find some fine specimens of information attitude when you snoop around your fileservers or shared drives. Just as our inboxes say a lot about us, our folders say a lot about our company's information strategy.

WHERE'S YOUR ENTRY POINT?

One of the strange things that we seem to forget in dealing with information is that how we store things isn't necessarily the same as how we retrieve things. What I mean is that when we create information and want to publish it, we have to 'park' it somewhere. We therefore 'check' this information into an information management system. It's a bit like storing goods in a warehouse, or stocking the storeroom of a shop with procured goods. This is the 'supply' side of information.

The way users retrieve, search or find information in, for example a portal or website, is the other side of the information exchange.

This is the 'demand' side. It's the retail space or the shop floor.

The trick is to 'disconnect' the supply side from the demand side. Unfortunately too many people get this wrong. When they check-in information, they use the same information structure for the shop floor as they do for the storeroom, and that's simply not the most sensible way.

Stores don't work that way. Do you think supermarket aisles are replicated in the storage areas? No. In the storeroom, items are grouped together in a way that makes sense for the storeroom manager to work: big boxes with big boxes, liquids with liquids, cold with cold. But on the shop floor, things are grouped together in a way that is best for the customer. Barbecue sauce is close to the barbecue coals, which is near to the barbecue meat and the barbecue utensils.

The same applies to content. It's often a very good idea when designing an information system to consider both areas of the company: what does the company shop floor look like (keeping the content consumer in mind), and how can we best organize the company storeroom (keeping the content suppliers in mind).

STRUCTURED OR UNSTRUCTURED?

We used to make a clear distinction between structured and unstructured information. Structured information was basically data: what you could put in a database. The nice thing about structured information was you could sort it. Records in a database, numbers in a data warehouse – it could all be easily manipulated because we'd stored it in a structured way.

Unstructured documents were all the rest: documents, PowerPoints, PDFs, drawings. We could store this unstructured information, but that was all we could do. We couldn't sort or manipulate it. All we could do is put it somewhere safe and then retrieve it later.

What was interesting was the difference between the growth rates of structured vs. unstructured information. Many companies had structured information growth rates that increased by as much as 30% annually: customers, addresses, accounts, revenue projections etc. Although 30% growth was a lot, it was still quite manageable in IT terms.

But unstructured information also exploded within our companies, with growth rates of well over 100% per year. This avalanche of information has grown out of proportion and is why our companies have so many information resources overflowing with (unstructured) data.

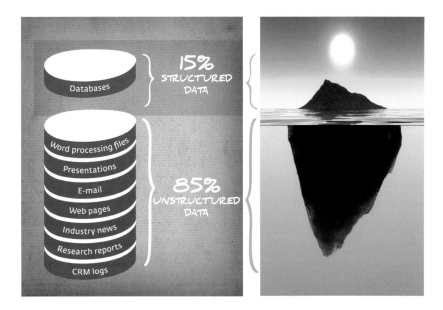

Today, we're beginning to realize that to make sound judgments, we need to look at both structured *and* unstructured data. If we want to understand an account for example, we not only want to look at the numbers (structured), but we also want to look at the contracts (unstructured), the email exchanges (unstructured), the chat sessions with this client (unstructured), and so forth. In order to define a comprehensive information strategy, we need to see 'the big picture'.

Nonetheless, the difference in growth rates still remains a challenge.

IT'S NOT INFORMATION OVERLOAD; IT'S FILTER FAILURE

In the end, information overload is not the chief problem. The real problem is that we don't have the right filters. Email is a good example. We've seen our daily ration of emails grow steadily over the last years to the point that it has now become a real burden for many of us. But we still have horrible filters. Most people only have one filter on their email: the *spam* filter that is binary – an email is either good, or it's bad.

There are now companies (such as xobni) that offer much better filters, and actually will tell you which emails are relevant and which aren't. Give this technology a little more time, and they will improve and become more accurate, and will become extremely skilled in sorting the important emails from the less important.

So to recap, our challenge in Information Technology is not to produce even larger receptacles for information. Our challenge is not to implement faster and better storage systems. Our challenge lies in bringing intelligence into the information game. Our challenge is in building better filters.

But this means we have to rethink information altogether.

Today we have information systems that allow us to look at things like author, version number or title. But that is not what we really need.

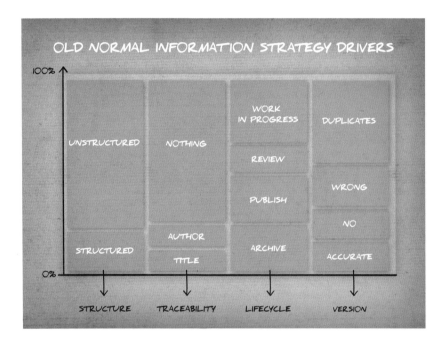

What we really need are 'clever' information systems that state what is 'relevant' for us, how good the 'quality' of document is, and who should read this document. But this demands that we work collectively towards that kind of information behavior. In the Old Normal, we had information systems that focused more on technology than on information.

In the New Normal, we will need to develop systems that focus on information, not on technology.

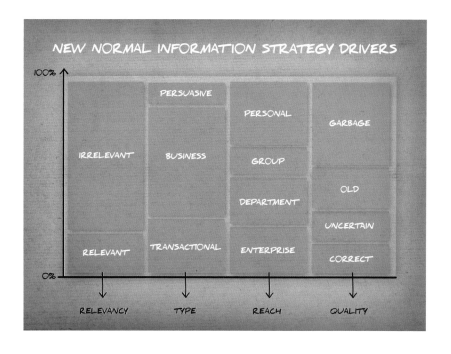

INFORMATION STRATEGY IS DYNAMIC

If you've cleverly thought out your information strategy drivers, I have some bad news for you. Information strategy is not a static thing any longer. It's becoming more and more dynamic as we move into the New Normal. In today's digital society, we can clearly see that because of consumerization, the information behavior of users changes faster than information systems do. Certainly faster than the information systems that we have within our companies.

Just take a look at the information behavior of young people entering your company today. For these digital natives coming into your organization:

Computers aren't technology. They are a part of their everyday life.

Multitasking is a way of life for them, and typing is always preferred to handwriting. For them, staying connected is essential. They won't tolerate the Internet being 'down', and they won't tolerate delays. If they experience delays on the network, they want to go home to work. Because they have much better Internet access at home anyway. But it's not just the younger generation; everyone's information behavior is shifting because of the New Normal.

The gap between information behavior and information systems becomes even more apparent when you take a look at the tools that people now use at home, and what they use at work. We now see that some information behavior at home, for which we use tools 'in the cloud', don't even have a counterpart in the office anymore.

Take email. Many corporate email systems have limitations on very large attachments, or on the amount of emails that you can store. That's why many people have started using systems like Gmail from Google.

While many corporate email systems have capacity limitations, the solutions like Gmail virtually do not. We will use tools like YouSendIt.com and WeTransfer.com to send really large files (like videos, or pictures), and while we are used to using this at home, there are often no corporate systems that have similar functions. We will use tools like Facebook for personal interaction, and Twitter for alerting and broadcasting information, often to find that no similar type of corporate systems exist.

And this is a fundamental shift in the New Normal. Before, IT departments developed corporate systems, and educated the users how to work with them. In the New Normal, we pick up new information behavior patterns at home, get acquainted with these new tools at home, and them bring them to work. And educate the IT departments on these shiny new tools.

INFORMATION BEHAVIOR

HOW DO YOU DEAL WITH...	PERSONAL	COMPANY
EMAIL OVERLOAD?	XOBNI	?
EMAIL ARCHIVING?	GMAIL	?
PERSONAL INTERACTION?	FACEBOOK	?
DOCUMENT SHARING?	GOOGLE DOCS	?
ALERTING?	TWITTER	?
INVITING?	DOODLE	?
LARGE FILES?	YOUSENDIT	?

BUT WHO?

So again the question is, *who* should do this? Is this a business problem, to be solved by management? Or is this an IT problem, to be tackled by the IT department?

As I described in my previous book, I believe the days of IT departments are numbered, at least in their current incarnation. I believe in a resurrection, a rebirth of the IT department, a revival of a new breed of professionals who can combine technical thinking with business thinking.

This is exactly what is needed in this space of information strategy as well. We need clever generalists who understand technology, understand the consumer of information and their usage patterns, and who understand the content that is relevant to their business. These are people or teams with multifaceted views and multidisciplinary skills.

These hybrid skill sets will make all the difference for whether companies can actually build an information strategy for the New Normal.

And you don't just hire these people. You breed these people. You cultivate, nurture and grow these people. Can you build the right team to turn your company into a survivor in the New Normal information age?

CONCLUSION

As we move towards the New Normal, I believe that many organizations, even the classic 'atoms' organizations, will have to move towards a quantum way of thinking about information. They will have to adopt an 'offensive' information attitude and will need to have an information strategy that enables people to solve 'mysteries' instead of giving them silos of information to solve 'puzzles'.

Information is about content, collaboration, intelligence and knowledge. Information is the cornerstone of our organizations. Defining an information strategy for the New Normal is crucial. In the New Normal, an information strategy centers on people. It's about how we work, share, publish, access and find information in the future.

There's no better way to end than to close with T. S. Eliot: "Where is the life we have lost in living? Where is the wisdom we have lost in knowledge? Where is the knowledge we have lost in information?"

T.S. Eliot, *The Rock*, 1934

DESCRIBING ORGANIZATIONS IN THE NEW NORMAL

"The achievements of an organization
are the results of the combined effort of
each individual."
— Vince Lombardi

"Organization charts and fancy titles
count for next to nothing."
— Colin Powell

"I won't belong to any organization
that would have me as a member."
— Groucho Marx

TOTAL ACCOUNTABILITY ZOOMS IN

Remember the third rule of the New Normal? The rule of total accountability. In Chapter 3, we talked about this rule in the context of the advertising industry, where the combination of data, metrics and real-time analysis brought transparency to a previously opaque industry.

Our argument is that in the New Normal, our ability to measure *everything* will result in total accountability, which will in turn lead to shared risk-taking in the form of increasingly blurred boundaries between companies. As the markets for services become more efficient in this progressively more transparent world, we would also expect a business to strike true win/win deals with partners, because in a measurable, quantifiable world, delivering win/win economics will be essential to sustaining a services business.

But total accountability will not just change the way we work with other businesses. It will change the way that we work with our own employees.

Consider the profundity of a statement by Marc Benioff, founder, chairman and CEO of Salesforce.com, at the recent 'Chatter' product launch in San Francisco: "I've run my whole company, for the last month, from my iPhone... I basically have the ability... to learn everything I need to learn about the company."

> Cloud 2 Launch:
> http://www.salesforce.com/
> video/events.jsp?v=
> xYs67Xgoq78

Salesforce.com has been an innovative force in the business world over the past decade, driven by Benioff and his intuitive understanding of the New Normal. As he wrote in a February 2010 TechCrunch post: "I quit my job at Oracle in 1999 because I couldn't stop thinking about a simple question: Why isn't all enterprise software like Amazon.com?" Built on this one key insight – that the user experience in the home had eclipsed the user experience at work – Salesforce.com created an entirely new market for enterprise cloud-computing solutions, and by early 2010 had over $1 billion in annual revenue and 72,000+ customers worldwide.

> www.salesforce.com

In his TechCrunch post, Benioff also revealed the next step in his thinking: "I've become obsessed with a new simple question: 'Why isn't all enterprise software like Facebook?'"

Chatter is a quintessential New Normal product. It is a work-life application poured into the mold of the most ubiquitous personal-life application on the planet. It is real time. And it brings an organization closer to the realm of total accountability.

'Chatter' by Salesforce.com

Imagine running your company from your mobile phone. Imagine reviewing performance at the division level, unit level, group level, project level, or even the individual level. Did Employee X close Deal Y yesterday? How was her recent presentation at Company Z received? How do her performance metrics compare to her peers? Imagine you are at the boarding gate in Heathrow, the Watergate in Washington DC or even at the Golden Gate Bridge in San Francisco. "Honey, stand over there and take a picture with the kids, the view is gorgeous (and yes, Jim just landed the biggest deal of the year!)."

Chatter is a leading-edge product today, but what about the leading-edge products of tomorrow? In the progression of data gathering and metrics, we have moved from cumbersome collection and slow reporting of basic, quantitative information – deals closed, accounts won, etc. – to real-time reporting of all conceivably-capturable events – emails sent, calls made, reports written, up-sells negotiated, etc. We can run some basic business analytics on quantitative information at the group level, but our ability to record and zoom is not yet matched by our ability to comprehend.

In the future, we can imagine the horsepower of data collection combined with the brainpower of sophisticated qualitative analytics. Employee data feeds on Chatter are interesting to management today because they never had access to this level of information before. Someday in the New Normal we will have the business intelligence capability to put all of these quali-tative inputs into a black box and return a single value: exactly how much value do *You* bring to the organization?

There is no 'back of the classroom' in the New Normal. There is nowhere that employees can hideout as wallflowers, watching their benefits accrue. If businesses play tug-of-war in the marketplace, we can now pinpoint every employee's position on the rope, check the calluses on their hands, the strain in their backs, and see their footprints in the dirt.

This all seems like good news for managers. After all, an employee is in one sense a 'service provider', and the transparency enabled by digital technolo-gies allow managers to select judiciously amongst service providers, just as companies are now able to base their selection of an advertising vendor on better information and performance metrics than in the days of Mad Men.

All this observation is not just to catch sub-standard behavior. By fol-lowing the progression of a deal – a successful closing or not – a manager can offer constructive feedback to an employee. Observation and meas-urement provide the basis for benchmarking, which can help to push and motivate an employee to perform at the peak of his potential.

Total Accountability is a system that avails itself to both carrots and sticks.

A BALANCING ACT

"The achievements of an organization are the results of the combined effort of each individual." If we believe Vince Lombardi's adage, then the modern corporation performs an astounding feat of human capital alchemy. After all, an ongoing entity pays out all of its salaries and bonuses each year to the individual employees who created value for it, and at the end still reports a profit. The unattributed 'the sum is greater than the parts' aptly captures the spirit of modern enterprise.

Using the new tools of transparency and total accountability, management will no doubt zoom in on underperformers in their organization and root them out. When we can understand the precise value that each employee brings to the company, one's compensation should never exceed one's value creation.

But here is where rule #4 – the loss of absolute control – comes into play. So far, we've talked about total accountability from the perspective of the employer. But if today management has the tools to pinpoint an employee's value to the company, tomorrow the employee will have access to those same tools. As in other areas of the New Normal, management will lose absolute control, and the information asymmetry it has about each employee's value to the company will vanish. We can expect the top performers to move up the pay scale or out of the organization. Quickly.

INTRAPRENEURSHIP

Retention of top talent has long been a management concern within organizations. In particular, companies that profit from innovation have experimented with inventive ways to keep innovators happy and productive within the larger entity. And we can quickly understand why retention matters. Imagine, for example, if PayPal had been able to retain all of its employees over the last decade: it would own YouTube, LinkedIn, Tesla Motors, Yelp, Yammer, Geni, Space X, and dozens of other companies.

The concept of 'intrapreneurship' – entrepreneurial activity within a corporation – was coined by Gifford and Elizabeth Pinchot in 1976 and further developed by Robert A. Burgelman of Stanford University. Burgelman was particularly interested in the tension between creativity and process within a corporation, writing about it in his 1983 dissertation: "Firms need both diversity and order to maintain their viability".

> Gifford and Elizabeth Pinchot, *Intrapreneuring: Why You Don't Have to Leave the Corporation to Become an Entrepreneur*, 1985
>
> Robert A. Burgelman, *Corporate Entrepreneurship and Strategic Management: Insights from a Process Study*, 1983

How does a company successfully transition from a start-up culture, where innovation is the sole engine of the business, to a mature company, where business processes and bureaucracy are essential for sustainability and scalability? How can a start-up become a mature company without stifling innovation, which can lead to a brain-drain that ultimately handicaps the company from recognizing the next big structural shift in its industry and keeping out ahead of the curve?

Companies like Apple, Microsoft and Intel have been working on this problem for decades. Andy Grove – co-founder and former CEO and Chairman of Intel – has collaborated with Professor Burgelman on several publications dealing with this topic, but Google's efforts on top talent retention may be the most instructive in the context of the New Normal.

GOOGLE'S GOLDEN RULES

If we could apply a single word to describe the digital revolution, it would be faster! That has applied not only to chip speeds, cost reduction and product cycles, but also to the growth of companies. Take Google's first four years as a public company compared to the first four years of rival technology companies Microsoft and Yahoo. In terms of both revenues and employees, Google's growth dwarfs its competitors.

In less than 15 years, the company has grown from two grad students in a Bay Area garage to 20,000+ employees in over 30 countries. This super-charged growth has given Google executives less time than normal to react to the tensions inherent in a move from a start-up culture to a corporate culture.

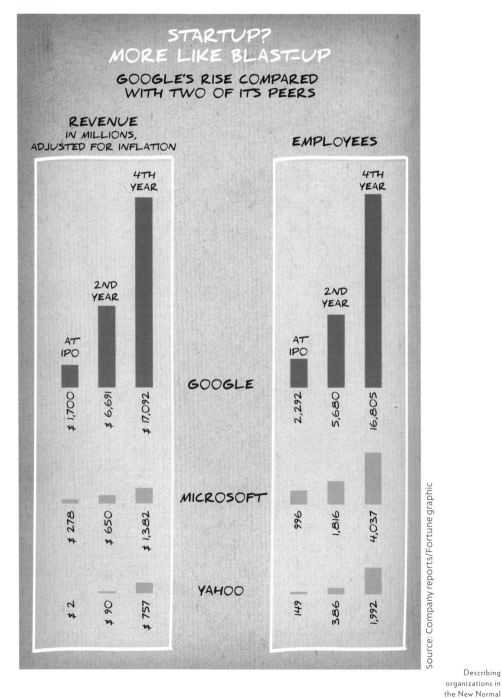

STARTUP?
MORE LIKE BLAST-UP

GOOGLE'S RISE COMPARED
WITH TWO OF ITS PEERS

REVENUE
IN MILLIONS,
ADJUSTED FOR INFLATION

EMPLOYEES

GOOGLE

4TH YEAR
2ND YEAR
AT IPO

Revenue: $1,700 · $6,691 · $17,092
Employees: 2,292 · 5,680 · 16,805

MICROSOFT

Revenue: $278 · $650 · $1,382
Employees: 996 · 1,816 · 4,037

YAHOO

Revenue: $2 · $90 · $757
Employees: 149 · 386 · 1,992

Source: Company reports/Fortune graphic

In December 2005, Google CEO Eric Schmidt and adviser Hal Varian published an article in Newsweek entitled 'Google: Ten Golden Rules', which started by stating: "Getting the most out of knowledge workers will be the key to business success for the next quarter century."

A few rules stand out as particularly interesting and 'New Normal' in nature:

www.msnbc.msn.com
search on:
Google: Ten Golden Rules

"HIRE BY COMMITTEE.

Virtually every person who interviews at Google talks to at least half-a-dozen interviewers, drawn from both management and potential colleagues."

This rule resonates with New Normal thinking. There is a de-emphasis on the control of management, and a related compression of corporate organization charts. There is also a community dialogue process around a candidate hiring, which is reminiscent of the community dialogue process around a product (discussed in Chapter 4). Google management doesn't tell its employees that they've found a great new colleague, but allows those employees to kick the tires themselves and then discuss.

"ENCOURAGE CREATIVITY.

Google engineers can spend up to 20 percent of their time on a project of their choice."

This is a particular model of intrapreneurship that has been used to strong effect at other institutions, including the US Military, where senior officials are encouraged to spend 20% of their time thinking about long-term strategic issues, and the 3M Corporation, where the 15% free time granted to engineers launched successful products including masking tape and Post-It notes.

Within an innovative company like Google, allowing employees to spend the equivalent of one day per week working on their own ideas acts as a pressure-release valve for creative types adjusting to a bureaucratizing corporate environment, and also leads to some interesting products: Orkut, created by Turkish programmer Orkut Büyükkökten during his 20% free time, is now the 61st most-trafficked website in the world according to Alexa, and boasts over 100 million active users.

www.orkut.com

"STRIVE TO REACH CONSENSUS.

Modern corporate mythology has the unique decision maker as hero. We adhere to the view that the 'many are smarter than the few', and solicit a broad base of views before reaching any decision. At Google, the role of the manager is that of an aggregator of viewpoints, not the dictator of decisions."

The wording here highlights a number of key New Normal concepts. The 'crowdsourcing' model of Wikipedia assumes a group of self-editing individuals can match experts in depth and surpass them in breadth. Similarly, Google's consensus approach prefers the balanced interests and viewpoints of the group to the elite 'decision-making' manager. Moreover, we again see a reshaping of the traditional power structure within the company, with the manager celebrated for aggregation instead of for exercising total control.

"DATA DRIVE DECISIONS.

At Google, almost every decision is based on quantitative analysis."

It is no surprise that Google takes a data-driven approach to management – after all, it was Larry Page and Sergey Brin's approach to extracting intelligence from data that launched their PageRank algorithm back in 1996. But it is interesting to see how widely Google is able to apply its approach in dealing with the pressing issue of talent retention.

In May 2009, a Wall Street Journal article revealed that Google was working on an algorithm to identify employees that are most likely to quit. Based on data from surveys, peer reviews, and promotion and pay histories, an early version of the algorithm was already able to identify employees who felt underused, a key complaint from those who end up leaving. As described by Laszlo Bock, Google's head of HR, the algorithm helps the company 'get inside people's heads even before they know they might leave'.

http://online.wsj.com

I recently talked with a Silicon Valley executive who spent his career in the semiconductor industry, first with Intel and later with fabless (fabricationless) semiconductor company Xilinx. In discussing the bell curve distribution of employee performance within a company, he said there were typically four groups: the low performers (5-10%), the average performers (40%), the above average performers (40%), and the top performers (10-15%). Within this distribution, the two management imperatives were to insure that the top performers are retained and fast-tracked, and that the bottom performers are 'managed out'.

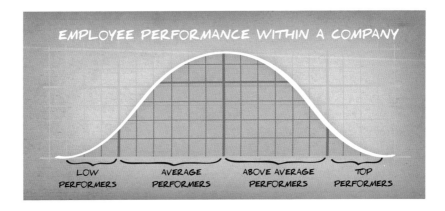

EMPLOYEE PERFORMANCE WITHIN A COMPANY

LOW PERFORMERS · AVERAGE PERFORMERS · ABOVE AVERAGE PERFORMERS · TOP PERFORMERS

In the New Normal, total accountability will help to quickly identify the bottom performers, but will also help the top performers understand their position and importance to the company.

Management will have to work harder than ever to retain top talent, leveraging intrapreneurship, crowdsourcing, and a data-driven approach that turns qualitative inputs into quantitative outputs to stay ahead of the curve on employee happiness.

INNOVATION AND OUTSOURCING

As we have seen, the New Normal will be filled with both challenges and opportunities relating to the 'managing up' and 'managing out' of employees. But what will the overall company look like in the New Normal?

Companies improve efficiency by concentrating on their comparative advantage in the marketplace. This is an extension of an economic tradition that goes back thousands of years, to tribal societies where tribe members were divided into skill-based subsets such as hunters, gatherers, fishers, builders and weapons makers.

During the Industrial Revolution and beyond, leading manufacturing companies relied on complex supply chains and a network of outsourced suppliers to produce parts that would ultimately be assembled and sold as a single product under a single brand. The Big Three American car companies embodied this strategy, operating as the gravitational center of a constellation of small suppliers that provided the components necessary for a Ford Mustang or a GM Cadillac.

The story of the semiconductor industry shows the impact of outsourcing taken one step further.

Semiconductor devices are electronic components that exploit the properties of semiconductor materials silicon and germanium. Semiconductor devices have replaced vacuum tubes in most applications. They are the foundation of modern electronics, including radio, computers, telephones, etc.

The industry had its roots in the invention of the transistor at AT&T's Bell Labs in the late 1940s and the invention of the integrated circuit in the late 1950's, but became a genuine industry in the 1960s, led by Fairchild Semiconductors, Texas Instruments, and, toward the end of the decade, Intel Corporation.

Because of the relative newness of the technologies involved, semiconductor companies in the 1960s had to invest in the entire manufacturing process, which included investments into material research, fabrication equipment, chemical science, etc. Accordingly, the industry was extremely capital-intensive, with heavy investments into both R&D and manufacturing.

By the 1980s, the standard industry model for semiconductors held that each company owned its own fabrication facility, which cost between $500 – $700 million to build, due to the complexity of the manufacturing process. Given this high barrier to entry, aspiring entrants to the industry had difficulties raising enough capital to both innovate in product technology and invest in manufacturing.

A couple of engineer entrepreneurs had a bold idea: why not outsource the fabrication of semiconductors to a low-cost partner and focus on innovation? This new outsourcing business model dramatically reduced the barriers to entry in the industry, and allowed a host of new innovators and entrepreneurs into the sector. These innovators in turn contributed greatly to the range of products within the industry.

In the last few decades, companies have moved beyond the outsourcing of manufacturing to the outsourcing of entire business functions. Today, for example, a fast-moving consumer goods company may outsource their payroll to ADP, their business applications to Salesforce.com, their IT business processes to Infosys, their logistics functions to DHL or UPS, and their business intelligence to IBM or HP.

THE BOUNDARIES OF THE NEW NORMAL

What, then, are the boundaries of New Normal organizations?

Remembering the Paradox of Theseus' ship may help guide our thinking. According to the Greek legend, Theseus' ship, upon its return to Athens following his battle with the deadly Minotaur, was preserved for generations. As an individual plank in the hull of the wooden ship rotted beyond repair, the Athenians would replace it with a new plank. Over time, the ship was renewed piece by piece, until one day there was not a single plank from the ship that Theseus had sailed. The question is if it still is Theseus' ship? And the larger question is, what constitutes the idea of a 'ship'?

Plutarch, *Life of Theseus*

Applying the same thought experiment to human beings, we can see that in society today, a person with a heart or kidney or liver replacement is still considered the same person, although some of his parts have been swapped out.

At the logical extreme, we could swap out arms and legs and vital organs and more, yet we'd still at some level feel that we were dealing with the same person. And so in the context of human beings, I think we'd answer that it is the 'soul' of the human that constitutes the idea of 'the human'.

So what constitutes the idea of a company? What is a company's 'soul'? The answer to this question will define our limit. I think we can then safely say:

LIMIT (OUTSOURCING) = THE COMPANY'S SOUL

THE SOUL OF A COMPANY

Going back to management guru Peter Drucker, "The business enterprise has two – and only two – basic functions: marketing and innovation. Marketing and innovation produce results; all the rest are costs." This echoes what we saw in the example of the semiconductor industry: focus on innovation, forget the rest. Without this approach, we wouldn't have Qualcomm, the largest fabricationless chip supplier in the world, and without Qualcomm we wouldn't have the pervasive mobile phone network that is a defining technology of the 21st century.

If everything other than marketing and innovation is a cost, then one day all remaining non-essential functions will be performed by the lowest-cost provider. The 'Software as a Service' (SaaS) and 'Platform as a Service' (PaaS) models pioneered by Salesforce.com point the way toward a future of increased efficiency and innovation potential for companies. If standard business applications are a cost, then the 'multi-tenancy', 'pay for what you use' approach can reduce fixed costs and deployment time for businesses, leaving them more time to focus on value creation.

In the New Normal, you should be asking yourself, "what is the 'soul' of my company?" The answer will depend industry to industry and even company to company, but outside of the soul or core of your company, nothing is sacred.

THE LIMIT OF AN ORGANIZATION

Remember the 'factors of production' from your old economics textbook? These were the resources employed to produce goods and services, and historically there were three key factors: land, capital goods and labor.

A fourth factor of production – entrepreneurial ability – was added to the mix later, and was defined as the act of combining the other three factors, making the decisions, and bearing the risk, in hope of generating an economic profit.

Being an entrepreneur has always been difficult. It involves a lot of responsibility and a lot of risk. But compare the process of starting a business 50 years ago vs. the process today.

—— **THEN**

Fifty years ago, you have an idea to invent a new and improved widget. You've got the basic design in your head, but you are not an industrial engineer, so you are going to have to recruit one. Searching for an engineer is going to be a time-consuming process, maybe you take on a secretary that can help with the HR recruiting process, and also to help you with your incorporation paperwork. You're not a trained accountant, so you are also going to have to pull a bookkeeper on board to keep your accounts straight.

You've finally found your engineer, and he's been able to build a nice prototype, so now you're off to make the rounds with the bankers to get some money to go into initial production. You've basically only got a few options for funding – the bankers that are in your town – so you end up with expensive terms, but at least you have a bit of money. You rent some space in a nearby warehouse, and hire a handful of workers to help execute on your first widget design while the engineer works on some additional modifications. You rent space across the street from the warehouse to set up your office, buy some capital equipment – phones, typewriters, desks, etc. – and hire some salespeople to start making sales calls around town. Once the orders start coming in, you'll have to buy a company truck to deliver your product.

Fast forward 50 years. You have an idea to invent a new and improved iPhone app. You've got the basic idea in your head, but you are not an app developer, so you post on elance.com to find one. Three hours later you have a dozen bids in your inbox. In the meantime, you incorporate the company online, and pick up the basic version of QuickBooks for $99 to handle accounting.

You've selected your developer, and he's been able to build a nice prototype, so now you're off to raise a bit of money to do the full development. You post some questions on TheFunded.com, an online community of 12,000 CEOs to research, rate, and review funding sources worldwide, and you get good feedback, which leads you to a solid funding source.

You hire a handful of coders to build your app while the developer works on some additional features. You buy $10 of Skype credit for international calls, stake out a good spot at a local Starbucks with free wi-fi, and you contract with a call center to handle customer service problems if your app has any bugs.

ENTREPRENEURS IN THE NEW NORMAL

LIMIT (ORGANIZATION) = ENTREPRENEUR

The New Normal is creating new opportunities for entrepreneurs. As a result of the outsourcing trend of the last decades, a broad spectrum of low-cost service providers now exist across most basic business functions, which means it is genuinely possible to run a scalable business by yourself.

Take the example of Adeo Ressi, a successful Silicon Valley entrepreneur. After expanding his worldwide rolodex of tech investors through his startup TheFunded.com, he recognized the opportunity to participate on the demand side by helping entrepreneurs to launch companies. This led him to create the Founder Institute – a hybrid model between a traditional tech incubator, an education program, and a networking organization – to empower founders to start meaningful and enduring technology companies.

Ressi essentially runs the Institute with a computer, a cell phone, connectivity to the cloud, and a small group of volunteers. To research potential 'Mentor' candidates (Mentors are tech CEOs who help guide the Founders through a basic curriculum), he uses LinkedIn. To circulate documents to Mentors in 15 cities throughout the world, he uses the 'share' feature of Google Docs. To plan the agenda for meetings for the Institute and get collaborative feedback, he uses Topicki. To organize Founder Showcase events, he uses Eventbrite. To host educational video content on the web, he may soon use Udemy, one of the early start-ups to launch out of the Founder Institute.

Does this one-man show mean that the Institute is a house of cards? Absolutely not. It has sponsorships from top Silicon Valley VCs, Mentor commitments from CEOs of the hottest start-up companies, an accomplished pool of applicants, and a growing list of successful graduate companies. What Ressi's example shows is the power of a new set of web-based tools.

All of these productivity tools have certain common traits that are quintessentially New Normal. They are easily set-up and used, they are interactive, they are web-based, and they are free or freemium. And all of these productivity tools enable the same thing. They allow the entrepreneur to exponentially leverage his time, reducing staffing needs and thus reducing fixed costs.

One of the most interesting proponents for outsourcing and leveraging time through productivity tools is Timothy Ferriss, who proposed his own form of limit in his 2007 best-selling self-help book.

Timothy Ferriss,
The 4-Hour Workweek:
Escape 9-5, Live Anywhere,
and Join the New Rich, 2007

Ferriss advocates 'lifestyle design', a concept meant to reexamine the career possibilities available to us in the 21st century. Although Ferriss argues that email, instant messaging and Internet-enabled PDAs complicate life rather than simplify it, his book is nonetheless premised on the broad opportunities unlocked by modern communication technology, which allows for the relatively seamless outsourcing of time-consuming work to low cost-of-labor countries.

LIMIT (WORKWEEK) = 4 HOURS

Ferriss' thesis is ultimately that the entry barriers to become a successful entrepreneur are lower than they have ever been, especially when the end goal is to earn enough money to live an enjoyable life rather than to accumulate bigger and bigger houses. In the New Normal, an individual with business savvy and connectivity should be able to achieve self-sufficiency.

THE COMPANY AND YOU

In post-World War II America, the social and societal compacts between and among individuals and companies were relatively straightforward:

① A (white) man had employment available for a wage

② A worker spent his career at one company

③ A man worked outside the home; his wife cooked, cleaned, and raised the kids

> Inspired in part by the work of Tammy Erickson, a business thinker focused on talent and innovation. www.tammyerickson.com

By the 1980s, compacts were breaking between all members and institutions in society.

Between the employer and employee, massive layoffs – fueled in part by manufacturing outsourcing – led to a fundamental break in the compact of a one-company career.

Where does the relationship between employers and employees now stand, and where is it headed in the New Normal?

One of the best thinkers on this topic is Tammy Erickson, a contributor to the Harvard Business Review (HBR) who focuses on talent, innovation and building a 21st century intelligent organization. In May 2010 HBR published an article entitled 'Restore Trust with Employees? Forget About It'.

> http://blogs.hbr.org/ erickson/2010/05/restore_ trust.html

Erickson summarizes the situation well:

"Trust in corporations was traditionally constructed in this way: The individual was loyal. The institution protected and cared for the individual. Employees professed to have no priorities outside their specific institution. And the corporation promised long-term opportunities and enhanced rewards for those who stayed...

We have been chipping away at one side of this relationship for decades, certainly since the extensive layoffs of the early 1980s. Very few, if any, organizations offer the option or expectation of lifetime employment, so even perfect loyalty doesn't necessarily result in protection and care anymore."

Given the accelerating pace of outsourcing across business functions outlined earlier in this chapter, it is no surprise that today's employees distrust their employers and any promises of job security or longevity. However, just as digital natives perceive and describe (digital) cameras differently than digital immigrants, so too will Generation Y workers view the employer/employee compact differently than Baby Boomers and even Gen X'ers.

Gen Y employees are now leading the move to the free-agent model. From the encompassing vision of the company as the provider of welfare and opportunity, we've shifted to a more limited and fragmented sense of what a company represents. While companies have evolved to become more shareholder-centric – all corporate actions should be in service of the mantra 'Increase shareholder value' – today's Gen Y employee is Me-Centric. He is focused on his own branding and employability. This new Gen Y employee stance establishes a more reasonable balance between the owners of investment capital and the suppliers of human capital, and will create a challenging but productive business environment for both parties.

> Generation Y, also known as the Millennial Generation, is generally marked by an increased use and familiarity with communications, media, and digital technologies.

GENERATION Y IN THE WORKPLACE

CHARACTERISTICS	IMPLICATIONS
INDEPENDENT AND LIBERTARIAN	– CAREER BELONGS TO INDIVIDUAL, NOT COMPANY – EXPECT TO BE TREATED INDIVIDUALLY – DEMAND FLEXIBLE SCHEDULES
RULE MORPHING	– EXPECT WORK TO FIT WITH OTHER LIFE COMMITMENTS – VALUE KNOWLEDGE AND SKILL, NOT TENURE
TECHNOLOGICALLY PROFICIENT	– DEMAND UP-TO-DATE TECHNOLOGY – MANAGE THEIR OWN INFORMATION AND COMMUNICATION
TRIBAL AND NETWORK CENTRIC	– WANT SOCIABLE WORKPLACES – 'WHERE I LIVE' IS MORE IMPORTANT THAN 'WHERE I WORK' – RELUCTANT TO RELOCATE – VALUE TEAMWORK – VALUE BELONGING TO CORPORATE COMMUNITY AND CONNECTING WITH A MENTOR
ACCUSTOMED TO FREQUENT MARKETING MESSAGES	– WANT FREQUENT FEEDBACK – WANT TO BE HEARD

Concours Group Presentation, by Tammy Erickson

So what is the New Normal compact between employees and employers? As Tammy Erickson puts it, "The organization will provide interesting and challenging work. The individual will invest discretionary effort in the task and produce relevant results. When one or both sides of this equation are no longer possible (for whatever reasons) the relationship will end."

The Gen Y crowd will also chafe under excessive control and bureaucracy. We are moving from 'command and control' to 'coordinate and cultivate', and this means a move from vertical orientations to horizontal orientations. Remember the 11-layer org chart at Alcatel at the start of my career? Gone. Most large organizations now work with 6-7 layers, and world-class organizations are targeting as low as 4 layers. Some innovative companies have replaced the top down structure completely, looking at new organizational models.

Cisco, for example, has moved from a traditional org chart to a system of councils and committees made up of executives from across the company. These teams provide strategic advice and evaluate the progress of various projects, and as of July 2009, there were 59 internal standing committees.

http://www.networkworld.com/community/node/44218?page=1

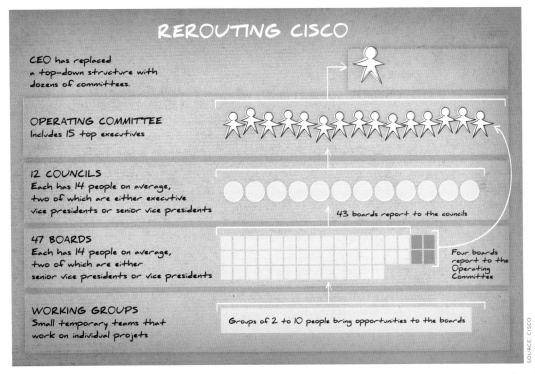

REROUTING CISCO

CEO has replaced a top-down structure with dozens of committees.

OPERATING COMMITTEE
Includes 15 top executives

12 COUNCILS
Each has 14 people on average, two of which are either executive vice presidents or senior vice presidents

43 boards report to the councils

47 BOARDS
Each has 14 people on average, two of which are either senior vice presidents or vice presidents

Four boards report to the Operating Committee

WORKING GROUPS
Small temporary teams that work on individual projects

Groups of 2 to 10 people bring opportunities to the boards

SOURCE: CISCO

T-SHAPED INDIVIDUALS

If you are a large organization, rotating your capable young employees through different business functions is probably the only way to engage and retain them, but an institutionalized rotation program also gives you another important benefit; it helps create T-shaped individuals.

The concept of the T-shaped individual was popularized by Ideo, a world-renowned design company based in Palo Alto, California, that has been ranked in the top 25 most innovative companies by BusinessWeek and does consulting work for the other 24 companies in the top 25. The "T" is meant to represent the breadth and depth of skill sets across business functions. A T-shaped individual has both a deep knowledge in a specific discipline as well as a breadth of skills in different areas.

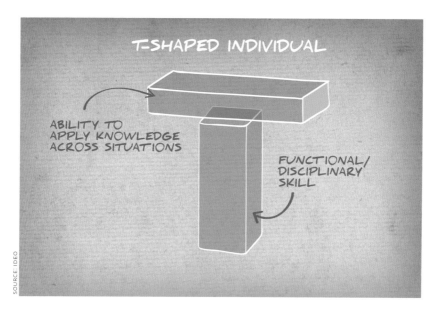

T-shaped individuals are contrasted with I-shaped individuals, who have very deep domain expertise but without a broad base of knowledge. Without the breadth of skills, an I-shaped individual is often unable to implement his expertise in the real world. You can think of it in terms of two shapes of nails: a straight nail shot into a wall might go straight through it, but a T-shaped nail will reach out to grab the wall.

Tim Brown, the CEO of Ideo, framed his thinking on I-shaped individuals in this way:

"When you bring people together to work on the same problem, if all they have are those individual skills – if they are I-shaped – it's very hard for them to collaborate. What tends to happen is that each individual discipline represents its own point of view. It basically becomes a negotiation at the table as to whose point of view wins, and that's when you get gray compromises where the best you can achieve is the lowest common denominator between all points of view. The results are never spectacular but at best average."

Morten T. Hansen,
T-Shaped Stars: The Backbone of IDEO's Collaborative Culture,
An Interview with IDEO CEO Tim Brown,
ChiefExecutive.net

I argued in my previous book that IT specialists have to broaden their basic knowledge of the overall business picture to remain relevant. I believe that in the New Normal, the same applies for a business specialist; that is, a marketing expert has to have a broad understanding of his entire business, even the IT function.

We have too many people who are I-shaped;
we need to start crossing the T's.

CONCLUSION

In the New Normal, organizations and the people within them will face a host of competing pressures:

If you are a bad worker, you will have nowhere to hide. Increasing accountability and transparency, driven by technologies, will shine a light on every corner of the company. In the New Normal, you are either performing, or you are gone.

If you are a top worker, the world looks much different. Your company will scramble to engage and retain you. Meanwhile, you'll have low barriers to entry if you want to explore other options, or start your own venture.

The form of a company will also change, as all functions other than marketing and innovation are increasingly sourced out to specialized service providers. Even core staff will be providing 'Labor as a Service', with a new and more dynamic compact governing the relationship between employees and employers.

With more transparency and more dynamic labor arrangements, we can expect more honest internal appraisals of companies and their culture. Will this drive efficiency and innovation? We'll know soon enough.

INNOVATION IN
THE NEW NORMAL

"Innovation is the central issue
in economic prosperity."
— Michael Porter

"Learning and innovation go hand in hand.
The arrogance of success is to think
that what you did yesterday will be sufficient
for tomorrow."
— William Pollard

"Innovation is the process of turning ideas
into manufacturable and marketable form."
— Watts Humphrey

THE SEEDS OF INNOVATION

Innovation has always been essential to business. In the New Normal there will be new factors to consider when calibrating your organization for maximum innovation, but first let's take a historical look at the process of bringing something *new* to market.

The conception of a new idea is mystical. How does one create something truly new out of things that already exist?

The moment of conception finds its spark beyond rationality. Sir Isaac Newton was just Isaac Newton until a falling apple plunked his head and created the idea of calculus. Percy Spencer invented the microwave after a candy bar melted in his pocket while he was working with radiation. Archimedes captured in a word the moment of discovery – Eureka! – after thinking about King Hiero's gold crown dilemma in the bathtub. Einstein, who made some of the largest leaps from old thoughts to new ones, concluded: "Innovation is not the product of logical thought, although the result is tied to logical structure."

For inventor Edwin Land, the moment of inspiration came from his young daughter's innocent but insistent question on a driving trip home from the Grand Canyon in 1944. Referring to the photographs that her father had spent all afternoon taking, Jennifer Land asked simply: "Why can't I see them *now*?"

Surely the question had been asked before of other camera-happy parents returning from family trips. But Edwin Land was particularly well positioned to find a satisfying answer for his daughter.

Dr. Land, as his friends and colleagues called him, was a college dropout from Harvard – perhaps the first in a great tradition of tech innovators – who invented inexpensive filters capable of polarizing light. He established the Land-Wheelwright Laboratories in 1932 to commercialize his technology.

By the time Jennifer asked her question in 1944, Land-Wheelwright Laboratories had been renamed the Polaroid Corporation, and the polarizing technology had been used in a range of products from sunglasses and color animation to passively-guided smart bombs in World War II. But it was this trip to the Grand Canyon that set Edwin Land to the task of inventing an instant camera and the associated film.

Land was idiosyncratic in ways that made him ideal for the role of innovator. He was notorious for marathon working sessions, in which he would experiment and tinker until a problem was fully solved. During these sessions, he needed to be reminded to eat. He once stayed in the same clothes for eighteen straight days when working on the commercial production of polarizing film, and eventually he transitioned to separate teams of assistants working in shifts at his side – when one team tired, the next was brought in to continue the work.

The relentless work paid off – just three years after Jennifer Land's question, "Why can't I see them *now*", Dr. Land had his answer: "You can."

The Instant Camera was an instant hit, with demand far outstripping supply over the first few product releases. Many imitators, including Konica, Minolta and Kodak, tried to compete in the new marketplace, but the Polaroid Corporation dominated the industry at the outset and maintained its position as a market leader over five decades, with sales peaking at $2.15 billion in 1992.

AND THEN CAME DIGITAL

Vandana Singh and Amy Sonpal summarize the course of events succinctly in their 2007 case study on the downfall of Polaroid: "In late 1980s, digital technology revolutionized the picture taking industry. Polaroid, under the influence of its immensely valued core business of instant photography, could not anticipate the magnitude of challenges from digital evolution."

> Vandana Singh and Amy Sonpal, *The Downfall of Polaroid: Corporate Lessons (Part A)*, 2007

The point here is not that digital came in and shook up the analog world, though it certainly did. The point is that the dominant market leader in instant photography failed to understand the next innovation in instant photography until after its competitors had taken a significant head start. Fuji and Dycam brought digital cameras to market in the late 1980s, while Polaroid's first digital release was in 1996. Polaroid was 'under the influence' of its core business, and thus missed an opportunity to convert its market advantage in one sector into a market advantage in another.

INVENTIONS AND INNOVATIONS

The story of the Polaroid Corporation and instant photography is instructive and straightforward: a single innovation can create a multi-billion dollar industry, but it cannot hold off the creative destruction of the next innovation. A company that innovates to create an industry will almost certainly be a market leader, but it can't maintain this position by momentum alone.

The story of the innovation in instant photography that ultimately sunk Polaroid – the birth of digital photography – is equally fascinating in its implications for innovation strategies in the New Normal.

The digital camera is celebrating its 35th birthday in 2010, which should be a cause for head-scratching given our observation a few paragraphs ago that Fuji and Dycam first introduced the digital camera in the late 1980s. While the digital camera was brought to market in the late 1980s, it was actually invented in 1975 by a young engineer working at the Eastman Kodak Company named Steven Sasson.

Sasson had recently earned his master's degree in electrical engineering when his supervisor, Gareth A. Lloyd, asked him to work on a broad assignment: build a camera using solid-state imagers (in non-technical speak, a digital camera). A few years earlier, Texas Instruments had developed an electronic, filmless camera, but it utilized analog electronics.

By December 1975, Sasson had a prototype that was ready for use. It weighed eight pounds, and its first picture – of one of his lab assistants – was a black-and-white image captured at a resolution of .01 megapixels. It took 23 seconds to record onto a digital cassette tape and another 23 seconds to read off a playback unit onto a television. For comparison, today's sophisticated mobile phones capture images at 1,000× the resolution at 5,000,000× the speed while weighing 40× less.

By 1978 Sasson and Lloyd had been granted a patent for the digital camera, and yet it was only in 2001 that Kodak began selling mass-market digital cameras. So why did Kodak wait so long to bring this key innovation to market?

Kodak – like Polaroid – definitely fell under the influence of its tremendously successful photographic film business, but the effect on its decision-making was slightly different. In the case of Polaroid, it simply overlooked the potential impact of digital photography on its business and thus failed to invest sufficiently in the product and marketing cycle.

Kodak management, on the other hand, saw the potential for digital photography early on, but subsequently came to view the product as competitive to its core revenue generator of photographic film.

A 2009 Wired magazine article on the internal politics of the digital camera at Kodak summarizes the history well:

Mic Wright, *Kodak develops: A film giant's self-reinvention*, Wired UK, March 2009

> "The innovation failed to gain backers in the then sprawling company. 'Some people talked about reasons it would never happen, while others looked at it and realized it was important.
>
> Kodak's executives were not enthusiastic... Early objections were intellectual, but that changed: 'By the late 80s they were coming from the gut: a realization that [digital] would change everything' -and threaten the company's entire film-based business model...
>
> Kodak's reluctance to let go of the vast profits from film was understandable. In 1999, its film sales rose by 6.5 percent to $3.1bn. Todd Gustavson, curator of technology at the George Eastman House museum in Rochester, New York, says: "Kodak was almost recession-proof until the rise of digital. A film-coating machine was like a device that printed money."

Rather than revolutionizing the industry and becoming a market leader in the new paradigm, Kodak chose instead to build up its intellectual property portfolio relating to digital photography while continuing to push its traditional film business.

So was this a savvy market strategy or a serious management blunder?

Some observers have defended Kodak's decisions, claiming that they would have needlessly cannibalized their own market share by introducing a revolutionary new product offering. An old Western saying goes, "The pioneers take the arrows, the settlers take the land." In a business context, we can understand this to mean that the first-movers don't always have the advantage, and that later market entrants can avoid mistakes and leverage brand name to catch up. In the case of Kodak, the adage seems to have proven true. In 2004 it leapfrogged Sony and Canon to become the leader in U.S. digital camera sales.

But most observers, including some company insiders, believe that Kodak survived by sheer luck, and just barely. In 2004, as digital was becoming a significant part of Kodak's revenue, CEO Antonio Perez initiated a restructuring that dramatically downsized Kodak's film factory workforce, focusing instead on boosting revenue from digital photography, specifically with a business-to-business focus. Referring to the 2008 economic downturn, which seriously impacted Kodak's revenue, Perez said, "If it had happened two years before, we would have been dead."

The final verdict on Kodak's specific digital strategy in the 1980s and 1990s is yet to be written, in part because 1,000+ digital-imaging patents from that time period are still being defended in court. As recently as January 2010, Kodak has initiated infringement suits – against Apple and Research in Motion for their use of Kodak imaging patents in the iPhone and Blackberry –that may one day lead to significant payouts. But overall, an emerging consensus argues in favor of the cannibalization of existing businesses for the sake of surviving an innovation cycle and suggests that Kodak's management was overly defensive in its decision-making.

Guy Kawasaki, an early employee at Apple who helped bring evangelism to the technology sector in the mid-1980s, summarized it best in a 2006 blog post intended to help companies navigate internal innovation:

http://blog.guykawasaki.com

> "**KILL THE CASH COWS**.
>
> This is the only acceptable perspective for both intrapreneurs and their upper management. Cash cows are wonderful, but they should be milked and killed, not sustained until – no pun intended – the cows come home. Truly brave companies understand that if they don't kill their cash cows, two guys in a garage will do it for them. Macintosh killed the Apple II: Do you think Apple would be around today if it tried to 'protect' the Apple II cash cow ad infinitum? The true purpose of cash cows is to fund new calves."

So far we've looked at two stories about the creative destruction unleashed in the photography industry by the rise of digital technology, and concluded that businesses need to spot the next wave of innovation before it capsizes them. Not too surprising!

But in the New Normal, we know that the waves of innovation will be larger and more frequent, because the barriers to entry will be lower, the markets will be bigger and more interconnected, and the speed of viral transmission will be faster. So how can we do a better job of staying ahead of the curve?

Revisiting perhaps the most-famous innovation case study – the Xerox PARC story – through a new lens might help uncover the beginning of a new innovation strategy for the New Normal.

THE PARC DILEMMA

This section is based on:
Henry Chesbrough,
Open Innovation:
The New Imperative for
Creating and Profiting
from Technology, 2003

The Xerox PARC story has been told many times. In the first act, a highly successful corporation has the insight to start a research laboratory – the Palo Alto Research Center (PARC) – that will help keep its technology fresh and innovative. In the second act, the research center is wildly successful, producing a number of paradigm-shifting technologies. In the third, tragic act (at least if you are a Xerox shareholder), we learn that very few of these technologies were put to use at Xerox, but were instead successful outside the confines of the company.

Let us not understate the importance and value of the technology invented at PARC.

Most of the elements of modern personal computing had their genesis at PARC, including computer-generated bitmap graphics, the Graphical User Interface featuring windows and icons, the WYSIWYG text editor, and the Ethernet.

In fact, when defending the similarities between Windows and the Macintosh OS from Steve Jobs' accusations of thievery, Bill Gates said, "Well, Steve, I think there's more than one way of looking at it. I think it's more like we both had this

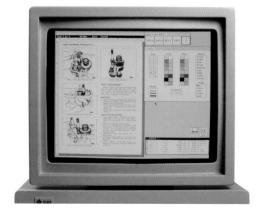

rich neighbor named Xerox and I broke into his house to steal the TV set and found out that you had already stolen it."

Jobs had toured the PARC facilities in 1979 and had been so impressed by the work being done on Graphical User Interfaces and human-computer interaction that he hired away several key PARC employees to work on the Lisa and Macintosh personal computers.

It is not just Apple and Microsoft that benefited from work that came out of PARC. Dozens of spin-off companies were created to commercialize Xerox's under-utilized technologies, including 10 companies that had IPOs, most notably 3Com, VLSI and Adobe. All told, hundreds of billions of dollars have been generated by inventions that Xerox PARC employees conceptualized, but which other firms successfully brought to market.

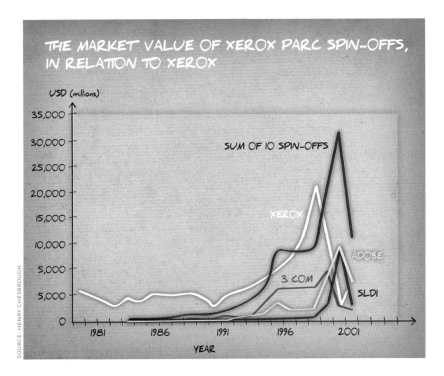

Many commentators have puzzled over the course of events that led to this outcome, but none have done so more cogently than Henry Chesbrough in his book 'Open Innovation: The New Imperative for Creating and Profiting from Technology'. Chesbrough uses the Xerox story as the quintessential parable of the digital revolution and its impact on business strategy. Let's take a closer look at what he uncovers.

Chesbrough starts by asking a logical question:

"How could a company that possessed the resources and vision to launch a brilliant research center – not to mention the patience to fund the center for more than thirty years, and the savvy to incorporate important technologies from it – let so many good ideas get away?"

To answer this, he first looks at some of the previous explanations that have been offered by other observers, but finds them unsatisfying. Yes, Xerox's corporate management on the East Coast probably did fail to properly value the technology being created at its West Coast research center. And yes there was probably some politics and infighting within the PARC facility that led to suboptimal decision-making. But for Chesbrough, these causes are secondary to the structural issues with the Xerox innovation formula.

As the Watts Humphrey quote at the beginning of this chapter suggests, innovation is a process that starts with invention and ends with a new product reaching the market. The researchers at PARC were extremely skilled at invention, but Chesbrough finds that the internal decision-making structure within Xerox management was not properly geared to match its invention engine, because it was operating in a 'Closed Innovation' paradigm.

CLOSED INNOVATION

Why is the Closed Innovation model that Xerox embraced, no longer adequate in the New Normal?

To trace the origins of Closed Innovation, Chesbrough looks back to the turn of the 20th century, finding a scientific community that was deeply skeptical of commercially directed research. Emanating from the European tradition of scholarship, this widespread attitude promoted the pursuit of science to discover pure ideas, not products. As a sign of the times, many scientists looked down on 'tinkerers' liked Thomas Edison.

During this period, there was also very little government-funded research in the United States – the federal government was much smaller at the time – and the university research system was in its infancy.

Given this state of affairs, it was necessary for large companies to build their own research and development laboratories to push the frontiers of their industries. These R&D facilities, which were expensive to set up and maintain, helped the leading companies pump out new products, which in turn allowed them to reinvest and grow their R&D functions. This created a virtuous cycle from the perspective of the leading companies, and these R&D centers became a barrier to entry for competitors.

The end product of the Closed Innovation model was dominant companies that were self-sufficient and self-reliant. This model generated mammoth technologies, which in turn created near-monopolies in telecommunications (AT&T), mainframes (IBM), copiers (Xerox), etc. Closed Innovation seemed here to stay.

Essentially, the research team would develop a kernel of an idea, and then the development team would either adopt the kernel and build a development schedule to bring it to market, or 'put it on the shelf' until the new idea fit more appropriately into a development schedule.

In the mindset of the New Normal, we are naturally suspicious of this point.

Time is not on our side as we head into the digital revolution.

A typical R&D company like Xerox was not well positioned to defend its shelved technologies, because in the past, there hadn't been other meaningful avenues for ideas to make it to market. They either succeeded in Xerox or they died in Xerox. Now, with hungry Venture Capitalists, an increased culture of job mobility, and a third new factor – the rise of sophisticated suppliers that dispatched the need for absolute vertical integration – start-ups became viable and attractive alternatives for frustrated research scientists and engineers.

WHAT IS VALUE?

But for Xerox, like other Closed Innovation companies, the single biggest factor in not defending its technologies was its inability to 'value' these technologies. After all, why spend precious resources developing a product and retaining key talent when the underlying technology doesn't appear to be valuable from your perspective?

And here we get to the crux of the matter: Closed Innovation companies perceive value differently than Open Innovation companies.

In the New Normal, we are in an Open Innovation paradigm.

If you are still operating in a Closed Innovation mindset, then you are expos-
ing your organization to the same mistakes that Xerox made in letting your
inventions fund somebody else's upside.

PLAYING CHESS VS. PLAYING POKER

Chesbrough frames the tension between Closed and Open Innovation as
the difference between playing chess and playing poker. For a Closed Innovation company like Xerox during the heyday of PARC, the R&D process was
biased toward products that fit into the existing Xerox businesses, and so
the management decision-making process was geared toward a closed-
system mindset.

Like a game of chess, the entire market was a closed system with a finite
number of possibilities, where customers, competitors and new products
interacted with one another in pursuit of the same objectives, and where
players shared a pretty clear view of the game.

A new technology aimed at a new market, on the other hand, had an
unknowable number of variables, and thus required a different strategy.
Like a game of poker, this scenario necessitated flexibility, quick decision-
making, and the nerve to test and probe the competitive landscape before
rapidly deciding to fold or double down.

Former IBM research director James McGroddy captures the distinction
perfectly in an interview with Chesbrough:

"When you're targeting your technology to your current business, it's like
a chess game. You know the pieces; you know what they can and cannot

do. You know what your competition is going to do, and you know what your customer needs from you in order to win the game. You can think out many moves in advance, and in fact, you have to, if you're going to win.

In a new market, you have to plan your technology entirely differently. You're not playing chess anymore; now you're playing poker. You don't know all the information in advance. Instead you have to decide whether to spend additional money to stay in the game to see the next card."

So while the industrial R&D companies were playing chess, Venture Capitalists started playing poker. With the poker mindset, these VCs were better able to recognize value in the shelved technologies within the R&D centers, and could properly align incentives for key talent to bring these technologies into a new market.

As for Xerox, it was caught in a perfect storm. Not only was it operating in a Closed Innovation paradigm, but its new inventions were also network-dependent by nature and thus needed a more open ecosystem to thrive.

The combination of these two factors meant that a lot of research at PARC which may have looked relatively valueless as a stand-alone product, especially in the existing universe of the Xerox business, became incredibly valuable once it left the Closed Innovation paradigm and migrated over to smaller companies and start-ups that cooperated with other firms out of necessity, if not calculated strategy.

FROM R&D TO C&D, CONNECT & DEVELOP

So what is a sustainable model if Closed Innovation is broken? What alternatives are leading companies utilizing to stay ahead of the competition, and what will define innovation norms in the New Normal?

Looking back at the general trends discussed early in the book – transparency, speed, fluidity – it is not surprising there has been a shift from Closed Innovation to Open Innovation. This means we no longer see so much vertical integration, but instead innovation is increasingly a managed process that takes place across the boundaries of companies. This change has been driven by all the factors previously cited, but it has also been enabled by our ubiquitous collaboration platform, the Internet.

The quintessential example is Procter & Gamble and its new Innovation model (and portal), P&G Connect & Develop.

P&G is an American consumer goods company founded in 1837, and is the fifth most profitable corporation in the world. P&G historically had the reputation of being an innovative company both in terms of its branding and its marketing (it actually produced and sponsored the first soap operas), but its R&D process over the past decades has been internal and closed, like many other leading enterprises.

In 2000, newly appointed CEO A.G. Lafley looked at P&G's innovation model – it was a Closed model but with a decentralized approach leveraging internal R&D from offices around the world – and recognized that it was not sufficient to meet the company's growth objectives. He challenged his executives to create a new model.

As Larry Huston and Nabil Sakkab, two senior executives at P&G, recounted in a 2006 Harvard Business Review article:

> Larry Huston and Nabil Sakkab, *P&G's New Innovation Model*, Harvard Business Review, 2006

"We knew that most of P&G's best innovations had come from connecting ideas across internal businesses. And after studying the performance of a small number of products we'd acquired beyond our own labs, we knew that external connections could produce highly profitable innovations, too. Betting that these connections were the key to future growth, Lafley made it our goal to acquire 50 percent of our innovations outside the company. The strategy wasn't to replace the capabilities of our 7,500 researchers and support staff, but to better leverage them. Half of our new products, Lafley said, would come *from* our own labs, and half would come *through* them."

P&G moved aggressively away from the Research & Develop paradigm and toward the Connect & Develop paradigm. Crunching the numbers, they realized that for every P&G researcher, there were approximately 200 scientists or engineers around the world that were just as good – which meant that a total of approximately 1.5 million researchers could become partners in the innovation process. The trick, Huston and Sakkab write, was to change a culture of Closed Innovation that had been in place for decades:

"We needed to move the company's attitude from resistance to innovations 'not invented here' to enthusiasm for those 'proudly found elsewhere'. And we needed to change how we defined, and perceived, our R&D organization — from 7,500 people inside to 7,500 *plus* 1.5 million outside, with a permeable boundary between them."

The model has been embraced and produced strong results. More than 35% of P&G products now have elements of outside influence in them, vs. 15% in 2000, and almost half of the initiatives in the product development pipeline contain key elements sourced outside the company's walls. R&D investment is now costing P&G less and making P&G more, and P&G's share price has doubled.

OPEN INNOVATION

In the New Normal, Open Innovation is imperative. Knowledge is highly decentralized, as evidenced by the growing number of patent applications across companies of all sizes and across the world. Savvy companies will tap into knowledge created and held by different players – universities, government research centers, independent experts, etc. – and combine it with internal knowledge. This mode favors the aggregator over the original content producer – a common theme in the New Normal – and speaks to the power of the platform.

Just as research will increasingly come from a variety of sources, so too will development.

In the heyday of Xerox, product development meant bringing in major customers to look at the new technologies coming out of the research department, but as we saw, this led to a bias against innovative products for unexplored markets.

Replacing the Closed model, development is increasingly coming from start-ups who are testing products in new markets with new customers. Many large companies now directly fund start-ups through internal venture capital funds, such as Intel Capital, or work with stand-alone venture capitalists that invest side-by-side with them. Contrary to the old days, when VCs were viewed as vultures stealing ideas from the idea warehouses of big internal R&D operations, VCs are now an accepted part of the complex knowledge landscape.

The May 2010 investment by the Toyota Motor Corporation into Tesla Motors is a recent example of this new attitude. As the President of Toyota, Akio Toyoda, explained:

"Through this partnership, by working together with a venture business such as Tesla, Toyota would like to learn from the challenging spirit, quick decision-making, and flexibility that Tesla has. Decades ago, Toyota was also born as a venture business. By partnering with Tesla, my hope is that all Toyota employees will recall that 'venture business spirit', and take on the challenges of the future."

> www.toyota.co.jp

Analysts who followed the deal believe that Toyota's motivations go beyond the allure of the start-up mentality to Tesla's technology (there is a technology cross-licensing agreement in place between the two companies). Whether it's the technology or the spirit, Toyota is looking outside its own company walls to find speedy ways to bring better products to market.

It is also worth noting that innovation does not just refer to products, but to all aspects of a business. As John Seely Brown, Director Emeritus of Xerox PARC, says in the foreword of Professor Chesbrough's book, "We need to be innovative in the area of innovation itself."

P&G Director of Global Open Innovation Chris Thoen elaborated on this theme in a recent interview:

> www.youtube.com
> search on: Chris Thoen
> talks about P&G

"It's not only about finding the next technology; it's about finding the best overall design of a product; it's about getting the product in the best possible way to the consumer; and it really is about new business models and new channels to get the product to the consumer."

WHAT IS THE LIMIT?

Still, it is clear that in this new model of Open Innovation, we are not there yet. Many large companies continue to operate in terms of closed R&D, though those firms will quickly be left behind in the New Normal.

But even Procter & Gamble – who pioneered the shift from Research & Development to Connect & Develop – produces less than half of its innovation from external sources, even though the ratio of external vs. internal researchers is 200 to 1.

In the New Normal, we could expect this number to rise even further, to the extent that we might say that the 'Limit of Innovation' tends to Open, even Outsourced, Innovation.

LIMIT (INNOVATION) = ∞

Is this a heretical statement? Returning to Peter Drucker's notion that the only two core functions of a business are marketing and innovation, are we ready in the New Normal to abandon innovation as a core function and focus solely on marketing, which equals brand, which equals platform?

Probably not. Because of the openness of the New Normal, plain marketing will not do, because customers will understand and compare end products and thus constantly demand innovation.

The general trend is from 'thinking in solitude' to linking with and getting ideas from multiple sources of inspiration. Companies that can find great ways to bring new products to market (whether invented in-house or proudly found elsewhere), via *new channels and new business models*, will be the true winners in the New Normal.

LEVERAGING PLATFORMS

Platforms become the go-to-market for many organizations. Again, think about a company like Procter & Gamble.

Each of their 22 name brands is a platform for innovation, as is the parent company. In other words, if you have a consumer goods idea to sell, you should be interested in partnering with P&G whether 'you' in this case represents an individual engineer, a small-medium sized company, or a large company with an innovative product on your R&D 'shelf'.

If the leveragability of the P&G platform seems fairly obvious because 'Atoms' companies require a higher degree of execution expertise, think about financial institutions. Banks are a typical 'Bits' company, but here too we see the potential to leverage well-respected brands to bring 'proudly found elsewhere' payment innovations to the market.

Even the most 'Bits' of Bits companies, Facebook, spends most of its internal innovation time finding new ways to leverage its platform of over 500 million users to spur the creativity of an entire mini-industry of application developers. More than half a million active apps now run on the Facebook platform, and each new app perpetuates a virtuous cycle where more new users are interested in joining the community because there is more functionality, and in turn more new developers join the community because there is an increasing user base.

Today more than 1,000,000 developers and entrepreneurs contribute to the Facebook platform, which is similar to the number of researchers that work with P&G through P&G Connect & Develop.

USER ADOPTION

As we enter the New Normal, we enter a new paradigm in which we can get instant access to a very large population. A new song on iTunes reaches crowds all over the globe. We also enter the realm of super-charged ('viral') user adoption.

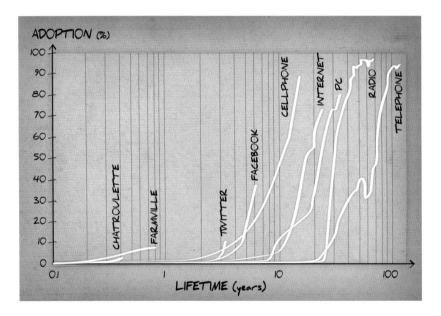

The graph above shows the rate of adoption of new technologies in the US (it is an extension of a Forbes graph with some new Internet players added in April 2010). In the first example chronologically, Alexander Graham Bell invented the telephone in 1876, but it took about 90 years before 90% of American households had one in their house. You can see later technologies enjoying ever-increasing adoption rates, from the radio to the PC to the Internet and the cellphone, and even to Facebook and Twitter.

Going back to the concept of the tiered ecosystem in the New Normal, we can separate the service providers (the YouTube content aggregator and Facebook ad consultancy) from the app developers. A consultancy cannot scale virally, and in this way is more like the suppliers from the auto industry; steady growth, with limitations. The app developers, on the other hand, are working with the purest scalable medium, and can achieve tremendous growth in increasingly compressed timeframes.

Take recent products like Farmville (a game on Facebook where you manage a virtual farm) and Chatroulette (virtual speed-dating that lets you set up a video call with strangers). Both benefit tremendously from the all-digital environment and can take months rather than years to reach a considerable adoption rate.

This is social marketing at its best. And we can imagine that as the "viralicity" of the platform and the applications get better, the speed of user adoption will get faster, to the point where it will tend to instantaneous.

LIMIT (USER ADOPTION) = 0

When a fun social game application like Farmville hits on the formula for super-charged adoption, it skyrockets out of the realm of small-tier player and becomes its own platform. There are now guidebooks, traders and offshore 'click farms' that feed off the Farmville ecosystem. So going back to our graph above, we now have the Farmville guidebook, which is built on top of the Farmville platform, which is built on top of the Facebook platform, which is built on top of the Internet platform. Like a stack of turtles all the way down.

And as far as striking gold? Zynga, the social gaming company founded in 2007 that counts Farmville as its crown jewel, has raised $366 million in Venture Capital in less than three years.

COLLABORATION

A second consequence of the New Normal in innovation, is how it should be approached. The New Normal considerably decreases user adoption time. Also, when you have an idea within this globalized economy, there is a good chance that somebody else has the same idea.

So, if you have a good idea, don't procrastinate. Deliver on it as soon as possible. If you miss a required competency, there typically is no time to build it in-house. Instead, look for possible partner companies that can contribute such competency. This should allow you to build the new product together, faster.

THE APPLE PARADOX

In the midst of all the trends we associate with the New Normal, one company that has not followed suit – (un)surprisingly – is Apple Inc.

Apple is undoubtedly one of the most innovative companies of the past ten years, and has come to embody the cutting edge of the digital revolution. Yet the management philosophies of Steve Jobs, its co-founder and current CEO, run counter to some of the basic tenets of the New Normal.

Apple scores low on openness, one of the key limits in the New Normal. But how does Apple score in Open Innovation?

> Leander Kahney, *How Apple Got Everything Right By Doing Everything Wrong*, Wired magazine, March 2008

Since 2008, there has been one key example of Apple leveraging its hardware platforms – the iPhone, iPod Touch, and most recently the iPad – to harness innovation and creativity from outside sources: the App Store.

The App Store has been a monster success, spawning numerous derivative app stores by other companies (although Apple tried to trademark the concept, suggesting that 'App' has the dual meaning of 'Apple' and 'Application'). As of June 2010, there were more than 225,000 applications available in the App store, and over five billion apps had been downloaded since the store launched on July 11, 2008.

But even this move towards a more open system is tempered by an app approval process, which gives Apple a final say over whether an application can be sold in the App Store or not. And because of the proprietary technical standards that Apple uses, an app that is not approved for the store cannot be used on any Apple device.

CONCLUSION

Innovation is powerful but it is not formulaic. Plenty of billion-dollar companies have been built on the back of a single innovation, but just as many companies have been downed by the next big shift that they missed or ignored. The question for market leaders today is:

How do we stay ahead tomorrow?

Tracing the history of managing innovation throughout the 20th and early 21st centuries helps to orient us in the New Normal.

Today, those companies that share, exchange, and collaborate will win, while those that stay trapped within their walls will lose.

For this chapter, however, the chief implication is that the relatively closed approach of Apple has for now surpassed the relatively open approach of Microsoft. (I use 'relatively open' in a particular context – obviously Microsoft has defended its intellectual property vigorously. The point is that they allowed their operating system to be widely licensed, as opposed to Apple, which tied its software to its hardware).

Going forward, however, will history repeat itself? That is, will the mobile phone space, like the PC space before it, ultimately reward the more open system?

Only time will tell, but given the most recent revelation that smart phones based on Google's Android mobile operating system (broadly licensed across multiple hardware makers) outsold Apple's iPhone in the US during the first quarter of 2010, I'm willing to bet that the trends of openness will carry the day in the New Normal.

CHAPTER

08

TECHNOLOGY STRATEGY FOR THE NEW NORMAL

Any sufficiently advanced technology
is indistinguishable from magic."
— Arthur C. Clarke

"The best way to predict the future
is to invent it."
— Alan Kay

"The future is already here.
It's just not evenly distributed yet."
— William Gibson

TECHNOLOGY STRATEGY FOR THE NEW NORMAL

It might seem a little odd to talk about 'Technology strategy' just as we reach the brink of the New Normal, where Technology stops being Technology and becomes 'normality'. But in my opinion, Technology strategy might be the most important chapter of this book. Companies that can re-think and re-position technology and re-create technology departments will have an enormous advantage over those that continue to hang on to the Old Normal way of thinking about technology and IT departments.

A BRIEF HISTORY OF IT

We use 'IT' as in IT department routinely today, but the term is relatively new. In the early days of digital technology, we actually called these departments the electronic data processing (EDP) department.

And that's what the department did: it processed data... electronically. EDP departments replaced pools of clerks pushing papers and compiling reports, gathering statistics and developing graphs, with big mainframe computers. The clickety-clack of typewriters gave over to the quiet hum of computing power harbored in the basement. Most of the data processed by EDP departments came from the finance department, so most EDP managers reported to the CFO. This explains why even today many CIOs still report to the CFO.

Time Magazine, Jan. 1983

For a long time, the 'technology department' was a central function that commanded powerful and expensive equipment and was run by technicians in white lab-coats with a 'guys from NASA' demeanor.

However things changed drastically in the 1980s with the advent of 'Personal' computing. The Apple II, and later the IBM PC and the Apple Macintosh, changed computing overnight. Gone were the days of million dollar mainframes in the basement and the 'elite' technologists who commanded these monsters of might. Suddenly, computing power was at everyone's disposal. In **January 1983**, Time magazine eschewed its annual 'Man of the Year' announcement and named the Personal Computer 'Machine of the Year'.

Bill Gates, the founder of Microsoft, coined the phrase "A computer on every desk", which was quite a revolutionary concept in the early eighties. Companies like IBM, which had become a global giant by supplying bulky mainframe computers, were completely caught off guard by the 'personal computing' revolution being driven by Apple, Atari and Commodore. Hastily playing catch up, IBM quickly slapped together its own personal computer and relied heavily on software from a tiny company in Albuquerque, New Mexico called Microsoft. A software giant was born.

The EDP department also stopped being the beating heart of an organization. The days of monster machines were gone now that everyone could have a computer. The department was replaced by a new concept: the Information Technology or 'IT' department. The wonderful new world of decentralized computing allowed people to get a handle on the processing of information, the most valuable of all corporate resources.

FROM BUILD, TO BUY, TO COMPOSE

But the techies, who had lost some of their status now that everyone had access to a computer, quickly found a new niche in order to remain vital to their corporations. With computing being distributed across a wide range of users, the power of the information itself became more apparent. Companies started envisioning how the information could be harnessed to drive profits and efficiency, and this in turn led to a new role for IT staff as architects in large-scale technology building projects. IT staff began 'building' information systems, constructing databases, and developing applications to help their companies be the most proficient at using information.

❶ BUILD

In the early days of Information Technology, the focus was on 'Build'. Programmers would create applications by custom building software and handcrafting databases and applications that could handle customers, accounts, payroll, general ledger, etc. In fact, you could compare this with the early days of car manufacturing – every model was a 'custom built' car, handcrafted and tailored to perfection. No two cars were the same. And no two software programs were either.

❷ BUY

In the second wave of Information Technology, the focus shifted from 'Build' to 'Buy'. In this era, we saw the rise of companies like SAP (the global market leader in ERP software), Siebel (the dominant player in CRM software, acquired by Oracle), and PeopleSoft (the leading player in HR software, also acquired by Oracle). These companies became global players by executing a simple business strategy; they persuaded customers that it was cheaper and better to buy software instead of writing it internally.

And these large software vendors were right on some level. There was a degree of standardization in each core software function – finance, logistics, CRM, HR – that could be usefully applied across a range of large companies. These large software vendors were able to hire the best coders to build the basic software and would then send out implementation teams to work on site with the customer's internal IT staff to tailor the core software to meet specific needs.

The value proposition was that companies could get more technology with less cost, less headache, and less hassle. The company could buy a 'pre-fab' software structure, make some modifications and be on its way. At least that was the promise.

ERP (short for Enterprise Resource Planning) is often seen as the 'IT heart' of an organization. This is the technology that helps a company plan its processes, optimize its resources and manage its finances. ERP implementation has been a major challenge for most companies that have embarked on the journey. I can always spot the ERP project managers in a full auditorium when I give a lecture. They're the disenchanted, dark, gloomy, chain-smoking types with bags under their eyes. I'm exaggerating, but most ERP implementations have been extremely difficult and have been compared to conducting 'open heart surgery' on a company.

In essence, you have two choices: you either adapt the ERP tool to your company, which means massive, painful and costly customizations, or you adapt your company's processes to the ERP tool you've chosen, which is equally complex and painful. ERP is in most cases a necessary evil, but has often produced an awkward win-win situation. Winners were the software vendor and the consultants doing the implementation.

Although the 'Buy' paradigm replaced the 'Build' paradigm, and virtually no company still 'builds' their own core software from scratch, the massive customizations laid on top of standard software packages essentially amounts to building by another name.

Following the car analogy, the days of custom built automobiles ended and companies switched to 'standard Model T' automobiles. However, the moment the Model T's left the factory, other companies started tuning and customizing their cars in such a way that you would hardly recognize it as a Model T anymore underneath all the modifications. It's as if the IT guys had a knack for 'pimping their ride' in such a way that every standard model became a tuned original.

❸ COMPOSE

Today, we're seeing another major shift. After the transition from Build to Buy, we're now witnessing the move from 'Buy' to *'Compose'*. Companies can now use the functionality of Internet-based applications which can be 'connected' together. In the ideal world, this means that companies can 'plug and play' applications that are run on the Internet and pay only for what they use. This is the Service as a Software (SaaS) model, and it is the third major transformation of Information Technology.

If we return to the car analogy, we've gone from building our own cars, to buying a car (and pimping it), to now renting a car only when we really need to get somewhere. This third transformation of technology, where applications will be provided over the Internet and run 'in the network', is what technologists call 'cloud computing'.

→ PYRAMIDS AND TENTS

If the car analogy doesn't suit you, try housing.

The old days of Information Technology could be compared to the construction of the pyramids in ancient Egypt. The famous pyramids in Giza were built 45 centuries ago to serve as monumental resting places for a ruling pharaoh.

The construction of the pyramids was an engineering miracle. It was also extremely labor-intensive. The largest pyramids were such monumental undertakings that they were truly the work of a lifetime – the first order of business for a young pharaoh was to get started on his pyramid, in the hope that it would be finished by the time of his death.

The first generation of Information Technology – the Build paradigm – was quite comparable. Companies constructed their information systems without the use of sophisticated tools, and thus demanded large amounts of manual coding (though these coders, unlike the slave laborers on the pyramids, were paid wages and subjected to safe if dreary working conditions). Early adopters of Information Technology like banks and insurance companies hired armies of software programmers to painstakingly build the desired information systems by writing software code one line at the time. The development of a new piece of software could take years to complete. Many CIOs didn't last long enough to see the completion of their pyramids.

In addition, companies found out that the upkeep of these monuments was extremely expensive. Yes, they took a fortune to build, and then companies discovered that they cost a fortune in upkeep as well, because these pyramids weren't built for power and prestige but for function and competitive advantage. If companies didn't invest in the maintenance of their newly created monuments, the information systems became digital ruins very quickly.

The Pyramid-building days in Information Technology are over. Most companies now have multiple ERP systems for finance, logistics, HR and CRM. But the upkeep of these grand legacy systems is not only expensive but also time-consuming. Because these systems are bespoke and 'built-to-last', they are also inflexible and difficult to adapt to new circumstances. Businesses see the value of agility in responding to opportunities in the marketplace.

In the New Normal, velocity trumps perfection. In the New Normal, there is a zero tolerance for digital failure, combined with an appetite for 'good enough' technology. Enter the Tents.

Business users don't want the IT department to build Pyramids any more when a 'Tent' is good enough technology. Tents are flexible, they are easy to set up, they can be deployed very quickly, and they are relatively inexpensive. Tents are practical and serve the purpose. You can choose which one you want based on your needs – capacity, functionality, durability – and you have a range of standard models to choose from. After a few years, you replace used tents with new ones.

For many IT people, setting up a tent, when you come from a time of constructing pyramids, seems degrading. This is a complex psychology that needs to be managed with extreme care.

THE NEW NORMAL CHANGES EVERYTHING

During the first few phases of Information Technology, the role of technologists has basically remained the same: technology follows business strategy, and facilitates the execution of the business game plan in the best way possible.

In the New Normal, technology stops being technology. In the old days of Information Technology, rivals competed to use new technology faster than their competitors, and barriers to entry in terms of money and time defined the contours of the competition. But in the New Normal, technology has become commoditized to the point that famed technology thinker Nicholas Carr wonders, "Does IT Matter?"

> Nicholas Carr, *IT Doesn't Matter*, Harvard Business Review, May 2003

When access to technology is equalized, technology becomes a commodity, the speed advantage is marginal and the barriers to entry are almost non-existent. Will your company be able to leapfrog its competitors because you were the first to install the latest version of Microsoft Office? Will your company beat the competition because you have the latest laptop model from Dell or HP? Did Mark Zuckerberg launch Facebook because he had a more powerful computer than the people at Friendster?

In the New Normal, companies need to start viewing innovation as the true enabler, and technology as the mean to drive innovation. In the last twenty years, becoming digital was a competitive advantage, and access to technology made the difference. In the next twenty years, in the New Normal, we have to focus on how to be clever with digital. Because being digital in and of itself will no longer define the winners and losers.

There will be no more 'technology projects', but all projects will be 'business projects with a technological angle'.

Instead of technology departments following the business strategy:

The role of the next generation of technology departments will be to lead the business into a world of technology-enabled innovation.

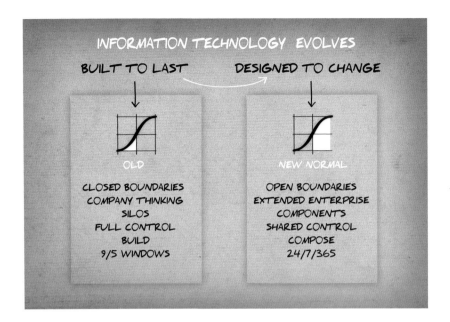

THE DEPARTMENT PREVIOUSLY KNOWN AS IT

In my previous book, 'Business/IT Fusion', I asserted that this move toward technology-enabled innovation should be a dramatic wake up call for IT professionals and a basis for drastically rethinking the old concept of the IT department.

In the New Normal, as technology stops being technology, the traditional role of the IT department disappears, the traditional role of the CIO disappears, and companies have to re-imagine the position and composition of the IT department. While the old IT department was a side-activity;

The new IT department must be a core asset.

I call this re-imagining process, the journey towards 'The department previously known as IT'.

The old IT department was primarily a butler to the business, trying its best to take orders and understand the 'demands' of the business partners. Its main function was the implementation of technological solutions. The new IT department will have to be the leader in digital innovation and lead the business towards the clever use of technology.

In a world that has gone digital, we can outsource the implementers, but we need to retain and develop the clever innovators who are well rooted in technology, but are fundamentally business thinkers, business leaders and innovators more than implementers.

The criteria above rules out roughly 80% of all IT departments that exist today.

I believe that most IT departments today are neither equipped, staffed nor positioned to be relevant in the New Normal. And the pace of change is speeding up.

Every day, new technology and new technology delivery mechanisms appear on the horizon and chip away at the traditional role of IT departments. Take 'cloud computing' as an example.

COMMODITIZATION AND THE CLOUD

Cloud computing is the next logical step in the evolution of Information Technology, and the central idea behind the cloud is an extension of the same trend that changed the economics of power generation toward the end of the 19th century.

This section is inspired by Nicholas Carr, *The Big Switch: Rewiring the World, from Edison to Google*, 2008

In the earliest days of power generation, direct current was the dominant paradigm, and direct current had two properties that favored a distributed generation system with large numbers of small generators located nearby their loads: non-standardized transmission voltages and inefficient transmission over long distances.

Given these limitations, almost all major industrial companies owned and operated their own power plants adjacent to their industrial operations, and those that had access to better, cheaper and more reliable power supplies translated the efficiency gain into lower product pricing, and thus had a market advantage over their competitors.

Building your own power supply was time-consuming, costly and surely necessitated some specialist staff to keep the boilers running at maximum efficiency. It was however a necessary part of the business in those days and also acted as a barrier to entry for smaller firms.

Things changed, however, when Nikola Tesla pioneered the alternating current, which allowed for far greater standardization of transmission lines and far greater distances between the generator and its load. The advent of alternating current soon led to the power grid. Power generation was no longer a differentiator between businesses but rather a commodity that enabled innovation. Sound familiar?

We can see the obvious parallels with the initial Build phase of Information Technology, and also with the commoditization of IT and the new move to cloud computing. Instead of each company buying its own hardware (servers), and installing its own software on these servers and employing a team of highly-trained technologists to implement and monitor and optimize this system, isn't it easier and more economical to access the desired functionality provided by a specialized provider 'in the cloud', which delivers these applications over the Internet? The answer is emphatically YES.

Cloud computing has been around for quite a while, and cloud-based email is an early example of a cloud product that gained widespread adoption among consumers. Just months after the web browser exploded onto the scene in 1995, we saw the use of email providers such as hotmail.com, which allowed you to use full email functionality via a web-browser instead of an email client. This also meant that your emails weren't stored on your servers but instead on servers 'in the cloud'. Hotmail became so popular that Microsoft acquired the company for $400 million in December 1997 (less than 18 months after the service was launched) as part of a defensive strategy to 'conquer' the Web.

Today, webmail services like Google's Gmail offer vast amounts of storage capacity, efficient and user-friendly interfaces, excellent search capabilities, and all sorts of functional extras like Gmail to Gmail chatting (Gchat), language translation, integration with calendars, etc. Gmail is free for individuals and extremely cheap for corporate users.

Webmail services offer a highly attractive value proposition, but have only made serious inroads in the business market over the past few years. Traditional email providers like Microsoft Outlook have argued to company executives that from both a security and reliability standpoint, their product offering is the only suitable option for business. In the New Normal framework, Microsoft is essentially saying that Gmail's 'good-enough' technology is good enough for the kids at home but not sufficient for serious enterprises. But the dangerous thing about 'good enough' technology is that it keeps getting better. And even 'near perfect' technology may let people down occasionally.

On the reliability front, Gmail has already surpassed most internal corporate mail systems. As the Wall Street Journal reported in September 2009, the research company Gartner estimated that corporate systems are available around 95% of the time, whereas Gmail guarantees 99.9% availability for its corporate customers.

The move towards webmail in a business context has not been a top-down process. While executives continue to wonder about reliability and security, corporate users have taken the situation into their own hands. So why do companies still have their own email systems? Why don't companies use these 'cloud based' services such as Gmail? Well, actually, a lot of users already do this. Many corporate users have both a corporate email account and a Gmail account, and the corporate accounts have more limitations – file size, capacity, etc. – than their Gmail account. As a result of these limitations, as well as a general fluency with their personal account, many corporate users actually redirect their corporate mail into Gmail because it is much more convenient.

This example illustrates the difference between the Old Normal and the New Normal, and between companies like Microsoft and Google. Microsoft has an Old Normal mentality, shaped during its heyday as a leader in the early phases of digitization. Redmond wanted a computer on every desktop, a laptop in every briefcase, and a server in every corporate basement. And Microsoft software on all of those machines.

OUTSIDE-IN, BOTTOM-UP

The leaders at Microsoft had a 'top-down', 'inside-out' approach: they believed that if they convinced executives to implement their products in the office, then employees would have to use them. And if employees used Outlook and Word at work, then they would continue doing so at home. Which they did. In the Old Normal.

But in the New Normal, this reverses. Google has the exact opposite strategy. Google believes that if they get people hooked on cloud based services at home, they will bring these tools to the work environment. And they do. People who use Gmail at home use it at work. People who use Google Docs at home use them at work. And eventually all of these rank-and-file users drive corporate executives to consider switching their entire system over to corporate Gmail accounts. Google in the New Normal has an 'outside-in', 'bottom-up' strategy.

These changes are unnerving to internal IT departments. In the Old Normal, IT had absolute control over what was installed on people's computers, what was allowed, and what was off limits. But in the New Normal, users know that as long as they have access to a browser, they can essentially do whatever they want, because their tools are in the cloud. Downloading quickly becomes a thing of the past.

More broadly, the shift toward cloud computing scares old IT departments just like the shift toward the power grid must have scared the power engineers in industrial companies at the turn of the 20th century. When computing power is as ubiquitous and easy to access as electricity, the role of the computer technician is diminished.

MULTI-TENANCY AND RETHINKING IT

Businesses are embracing cloud computing because it has the capacity to deliver new levels of the efficiency to IT processes through its multi-tenant architecture.

Think of a business like a small family and IT capacity like the house that the family lives in. Typically, the family only needs a small amount of living space, but because they will from time-to-time have a visit from the in-laws or a birthday party, they are forced to own more space than they need. This contingency planning means that most of the time, the family is under-utilizing its capacity, and spending more money than it needs to.

The multi-tenancy model allows the family to use space as needed, and pay for use. If the kids are out of town for the weekend, the family can reduce its living space, and if grandma comes to visit, it can increase its living space. This flexible architecture allows companies to reexamine their spending patterns.

Today, many IT departments have their IT capacity planned for the 'maximum potential capacity', but sometimes use only 20% or 30% of their capacity at any given time. Cloud computing allows companies to only plan for 'normal potential capacity', and use the cloud when necessary.

Today, many IT systems have been sized for the 'maximum number of potential users'. Very often only a fraction of these people use these systems at any given time. Cloud computing allows companies to rethink this, and only plan for 'normal usage', and use the Cloud when necessary as an 'on-demand' resource.

So the cloud is the next step in the evolution of computing power, and like all disruptive developments, it is perceived as a threat by old IT departments who become extremely possessive of 'their' systems, and 'their' servers.

But the threat of cloud computing to the status quo is also a catalyst for change.

IT departments are still populated with 'pyramid builders', deeply technical craftsmen who were instrumental in delivering the Old Normal in technology, but we don't need armies of technical craftsmen any longer in the New Normal.

What we need are people with technical backgrounds who can think as business leaders, and who focus on innovating with technology instead of implementing technology.

People who can drive a car, rather than build and maintain it.

FORGET BUDGETS. THINK INNOVATION PORTFOLIO.

Implementing a technology innovation portfolio is one of the best mechanisms with which to think about the potential of technology. Creating such a portfolio is what we call the 'Technology Innovation Sudoku'.

A classic way to look at investments is through the McKinsey framework of 'Run, Win, Change'. This is a simple way to break down a budget and classify the types of investments that are made.

Smarter IT investments,
The McKinsey Quarterly
Chart Focus Newsletter,
October 2007

In the 'Run' category, you put those investments that are necessary to stay in the game. These are 'keep the lights on' investments that won't generate a lot of value, but that you need to make in order to survive.

In the 'Win' category, you have those investments that will propel your company as a 'winner' in your field. These are the investments that will help differentiate your company and that can generate a (sustained) competitive advantage.

'Change' investments are pure innovation, where you try to generate significant (future) value. These investments, if successful, will 'change the rules of the game' altogether.

In a traditional IT budget, the breakdown across these three categories is roughly 70-20-10. The bulk of the annual IT budget is spent on 'keeping the lights on', while less than 20% is spent on projects that really make a competitive difference and less than 10% is spent on those super innovative projects with very high value creation potential.

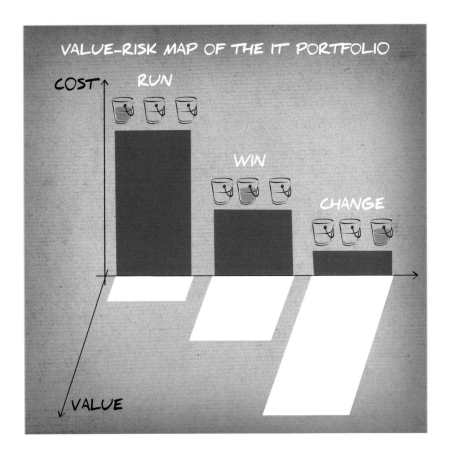

When you only look at the budget part, you're flying blind. When you only look at costs and risks, you are blind to the potential value creation of these projects.

It is only when you add the 'bottom' part of this equation that you begin to see the full picture, and that you have a 'value portfolio'. This allows you to move from making decisions on a COST+RISK basis to making decisions on a COST+RISK+VALUE basis.

ANOTHER DIMENSION

Now we need to move one step further. Although a three-axis framework pushes us to make better value-driven decisions *vis-à-vis* technology, it still lacks the complexity to allow us to see where these investments actually will make the difference. That's why it is vital to add another dimension to this story and talk about where these investments will be applied.

Based on thinking by Michael Porter, a classic pattern of focus for companies evolved around three major areas: Customer, Product and Operational excellence.

Michael Porter, *On Competition, Updated and Expanded Edition*, 2008

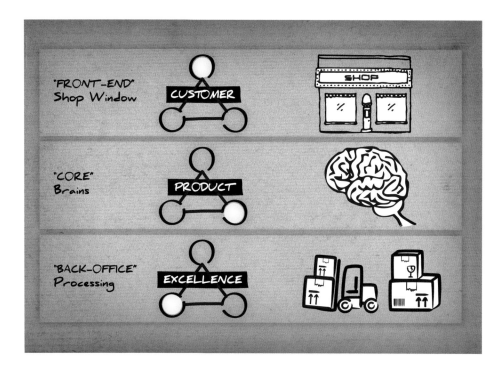

While we can certainly identify companies that have been extremely successful by focusing on one area – customer intimacy (Amazon), product differentiation (Apple), operational excellence (FedEx) – most companies won't choose a single category but will have a balanced view across these three focus areas. They will shift resources among these areas and determine their ideal mix.

When we map this to technological investments, we can see where the money goes.

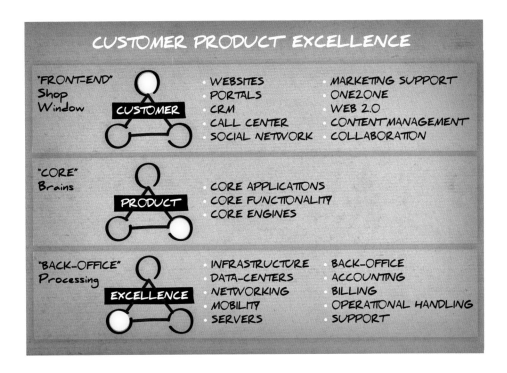

CUSTOMER PRODUCT EXCELLENCE

"FRONT-END" Shop Window — CUSTOMER
- WEBSITES
- PORTALS
- CRM
- CALL CENTER
- SOCIAL NETWORK
- MARKETING SUPPORT
- ONE2ONE
- WEB 2.0
- CONTENT MANAGEMENT
- COLLABORATION

"CORE" Brains — PRODUCT
- CORE APPLICATIONS
- CORE FUNCTIONALITY
- CORE ENGINES

"BACK-OFFICE" Processing — EXCELLENCE
- INFRASTRUCTURE
- DATA-CENTERS
- NETWORKING
- MOBILITY
- SERVERS
- BACK-OFFICE
- ACCOUNTING
- BILLING
- OPERATIONAL HANDLING
- SUPPORT

In the '**Customer**' bucket, we find all sorts of investments that will greatly enhance customer intimacy and the user experience aspect of dealing with your company. As more and more of the interactions with your customers become digital, digital interaction patterns will determine customer satisfaction, so this is an area of great interest. Here we find websites, portals, and CRM systems, but also the use of social networks, Web 2.0, and new delivery mechanisms such as Twitter.

In the '**Product**' bucket, we find offerings that relate to your core business. If you are an insurance company, this is the application set that calculates rates and premiums. If you are in the travel industry, this is the set of algorithms that searches for the best hotel rates, the best connecting flights, or the best package deals. Companies that deal more in 'atoms' than in 'bits' will have a harder time investing IT budgets into their core offering, but in general this bucket is the core set of applications that makes your company tick.

In the '**Operational Excellence**' bucket, we find all elements that ensure a flawless delivery on the technical side. These are often the things nobody 'sees' until they don't work. Here we have all those investments that ensure your recipe to counter the 'zero tolerance for digital failure' we talked about. And as long as we are not competing solely on the basis of 'operational excellence' this is the bucket that is most likely benefitting from the commoditization of IT services and the outsourcing of business functions to the cloud.

It is quite easy to map your technological investments into these three categories, but this is an exercise that IT and business have to do together in order to make the right choices.

		EXAMPLES		
		RUN	WIN	CHANGE
"FRONT-END" Shop Window	CUSTOMER	MAINTAIN WEBSITE	IMPLEMENT CUSTOMER PORTAL	ENGAGE IN SOCIAL NETWORKS
"CORE" Brains	PRODUCT	IMPLEMENT COMPLIANCE UPDATES	TRANSFORM MORTGAGE APPLICATIONS	DEVELOP A P2P LENDING RISK MANAGEMENT SYSTEM
"BACK-OFFICE" Processing	EXCELLENCE	UPGRADE NETWORK TO CONSOLIDATE DATACENTERS	MOVE DATA SERVICES INTO THE CLOUD	SUPPORT MOBILE 3G ACCESS TO SALES REPS

THE TECHNOLOGY INNOVATION SUDOKU

Things get exciting when we combine the 'Run, Win, Change' view of IT investments with the 'Customer, Product and Operational Excellence' view. This gives you a 'Sudoku' view of your innovation portfolio.

From left to right you move from the more traditional 'keep the lights' on projects towards the more innovative projects, and on the vertical axis you see where you put these innovations to work in your company strategy.

When you add up the numbers, they always sum to the same total. This is 100% of your technology innovation portfolio. But the discussion to have in your company is not "How much is 100% in dollar terms?" Rather, you should be asking, "How do we allocate investments in our IT portfolio to maximize benefits for our company?"

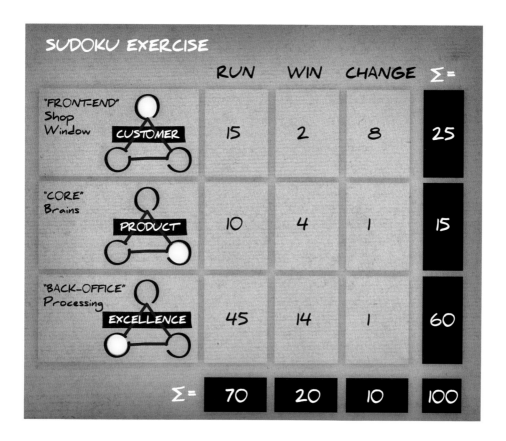

SUDOKU EXERCISE	RUN	WIN	CHANGE	Σ=
"FRONT-END" Shop Window — CUSTOMER	15	2	8	25
"CORE" Brains — PRODUCT	10	4	1	15
"BACK-OFFICE" Processing — EXCELLENCE	45	14	1	60
Σ=	70	20	10	100

CONCLUSION:
TOWARDS A NEW TECHNOLOGY INNOVATION ORGANIZATION

As the role of digital technology in our work lives has changed over time, so too has the role of our technologists and our approach to technology, innovation and competition within our businesses.

We have entered the era of cloud computing, where computation has become a commodity and is no longer a differentiator for businesses in and of itself. The waves of technological advancement will not decrease in the New Normal. On the contrary, the speed of technological revolutions will only increase. And therefore, velocity and agility become the cornerstones of your technological capacity.

Discussions about IT budgets become less relevant in the New Normal. Instead, the discussion should be about a company's innovation portfolio. You should steer the Technology Innovation Sudoku for the organization, and make the right trade-offs to ensure maximum value creation with your technological potential. Businesses will gain a competitive edge by being clever with technology. But who will lead the clever thinking about technology and its essential role as an innovation enabler?

Instinctively we should turn to our technologists, but the Business/IT alignment model is often broken beyond repair. The IT crowd is in a particularly demoralized mindset. From their original positions as the 'men from NASA' who powered their companies to new heights with their mysterious servers, they regressed to mere execution and now further still to basically clinging onto their jobs. With the advent of cloud services, in-house systems and servers are becoming a thing of the past, and ever-improving user interfaces and implementation processes make IT technicians an endangered species.

There is still hope for the department previously known as IT. After all, someone has to lead the thinking toward technology-enabled innovation, toward the deep integration of digital into our businesses, and it might as well be the techies. There is a spark in the IT crowd, but to ignite new creativity and innovation, we have to tear down the old IT structures and rebuild a new dynamic of thinking about and working with technology.

This requires extreme leadership on behalf of the CIO to dare to rethink his own organization. Only the bold and brave CIO will survive; those leaders who question the fundamentals of their own organization and then move a step further, to burn down their old IT department and see a Phoenix rise from the ashes, more powerful than ever before.

CHAPTER

09

THE BIG PICTURE

"We learn from experience that men
never learn anything from experience."
— George Bernard Shaw

"The sooner you fall behind,
the more time you'll have to catch up."
— Steven Wright

"My interest is in the future because
I am going to spend the rest of my life there."
— Charles F. Kettering

RECAP

We've covered quite a bit of territory in this book, so it makes sense to take a look back before we end with some thoughts beyond the New Normal.

In the opening chapters, we discussed a conceptual framework of Limits and Rules for understanding the New Normal.

LIMITS

Limits are used to put today's digital practices in perspective by comparing them to yesterday and projecting them into tomorrow and beyond. We found that:

→ **LIMIT (LENGTH) = 0**

We've gone from reports to memos to emails to tweets.

→ **LIMIT (DEPTH) = ∞**

With a browser window open to Wikipedia, we can become experts in any subject.

→ **LIMIT (PRICE) = ?**

Some say that price is moving toward Free. Others argue that the trajectory of price is far from certain. We'll wait and see.

→ **LIMIT (PATIENCE) = 1**

Consumers should have to provide their digital info once and only once.

→ **LIMIT (PRIVACY) = FISHBOWL**

Our lives become increasingly digital. Digital becomes increasingly searchable. Thus our lives become increasingly searchable.

→ **LIMIT (INTELLIGENCE) = REAL TIME**

We've gone from batch processing to real-time analytics.

We also looked at four key rules that will shape the New Normal:

Zero Tolerance for Digital Failure

A chair used to be a new piece of technology. Now it is standard. If a user sat in a chair today and it broke, he has reasons to be angry. In the New Normal, digital will be standard. In the New Normal, if a user's digital tools fail, he will be angry.

Good Enough Beats Perfect

The Flip Ultra digital video-recorder was inferior to other video-recorders on the market: it shot relatively low quality footage, had a minuscule view-ing screen, had no color adjustment features, no optical zoon and very basic controls. But it was small, inexpensive and easy to operate. And that was good enough to gain a market-leading position and get acquired by Cisco for $590 million in 2009.

The Era of Total Accountability

As consumers of products and services we are now better able to assess the efficacy of those products and services because of our access to real-time information and a widely searchable (one day perfectly searchable) digital record. This is good news for the consumer, but it raises the bar for producers.

Abandon Absolute Control

Org-charts are getting flatter. Wikipedia is bottom-up, and Facebook is side-to-side. Yes we can try to transfer hierarchies from old systems onto new ones, but we will be fighting against entropy. Also, machines are capa-ble of handling an increasing number of narrow tasks more efficiently than humans. The New Normal is a time for letting go.

Once we established the Limits and Rules framework, we explored specific strategies for customers, information, organizations, innovation and technology. All of these chapters are relevant from the consumer perspective. They are also aimed at specific business units, including marketing (customers), HR (organizations), R&D (innovation) and IT (information & technology). And of course understanding the trends in all of these areas is critical for management, present and future.

CUSTOMER STRATEGY

In the New Normal, the customer strategy needs to be built around *You*. This means abandoning the averages (2.3 kids and a white picket fence) that dominated the mass-media paradigm, and moving toward a fully targeted approach that treats each individual consumer uniquely. There are also several emerging patterns in the New Normal that companies should be aware of: the Long Tail (the opportunity to address niche markets), the Participators (people who have the capacity to contribute a great deal of valuable knowledge for free and already do on sites like Wikipedia), and the Community (digital communities that are discussing your product whether you want them to or not).

To get a customer strategy right, companies need to be leveraging digital – a post-bubble technology that is ripe for a Golden Age – putting out the right, customized content through a variety of sources, and then responding to customer demands in a personalized and *consistent* way across a number of channels.

INFORMATION STRATEGY

Building a cohesive information strategy is critical in the New Normal. The four pillars of information are content, collaboration, intelligence and knowledge, and the function of a complete information system is to build an effective value-add chain to turn the raw information into organizational knowledge.

In the digital paradigm we are generating and capturing absolutely massive amounts of information – text, pictures, audio, video – but many companies are not converting that raw information into a useful end product.

Most data that we are now amassing is unstructured vs. structured data. This is partly because we've spent a lot of time over the last 20 years focusing on technologies that enable us to create and capture all kinds of data.

In the New Normal, we need to be less focused on **T**echnology and more focused on **I**nformation: how can we create meaning and value from the information that we are capturing?

What will organizations look like in the New Normal? Our first observation is that the same pressures that result from Total Accountability – consumers hold suppliers to a higher quality level – also effect employees within a company. After all, what are employees other than suppliers of a service – their labor – to a company? This means that it is harder to be a wallflower in the New Normal. But it is also harder for companies to retain top talent, because those key employees understand their value to the firm, and have a wider pool of alternative employment options to choose from.

Overall, organizations will outsource more and more business functions until only its core remains; for most companies, this means marketing and innovation. Outsourcing means that the lines are blurred between organizations, as different discrete businesses perform various tasks to deliver a single product or service to a customer, and it also means that individual entrepreneurs or small teams will be able to execute quickly on a new business idea by tapping partners.

Outsourcing and mobility also changes the mindset of Generation Y workers, who increasingly view employment as a short-term trade of services for experience. This is an issue for HR and management to deal with in their quest to build efficient and innovative teams. Another is finding those T-shaped individuals with deep domain expertise and broad business understanding, who can be cross-functional and make sense out of the rapid churn of available business opportunities.

INNOVATION STRATEGY

Without an innovation strategy, companies will lose their competitive advantage in the New Normal. By looking back at the history of innovation and missed opportunities, we can identify a fundamental transition from Closed to Open Innovation, which helps to explain why innovation hubs like Xerox PARC failed to capitalize on their inventions while smaller start-ups were able to take them to market and succeed.

But open innovation is not only for resource-scarce start-ups. In fact, large companies like P&G are now leveraging their brand and execution prowess to bring more innovative products to market that were 'proudly found elsewhere'. This new 'Connect & Develop' model highlights a nuanced twist in the New Normal: value creation and innovation can happen at the product level, but it is perhaps even more potent at the channel and business model level.

Platforms are a potent value creation tool for large businesses in the New Normal, whether it is P&G's stable of 22 billion-dollar brands or Facebook's 500 million users. But in the pure bits realm, the increasing speed of viralicity and user adoption means that even small innovators can hit it big with a great application, with the potential to even create a new mini-platform.

Despite the trends toward open innovation, Apple's relatively closed eco-system means that it is not yet the end of history when it comes to innovation strategies. But with iPhone vs. Android shaping up to be a mobile version of the Mac vs. Windows showdown, we wonder if history will repeat itself.

TECHNOLOGY STRATEGY

Finally, we review a technology strategy that will help your company navigate through these turbulent times. The key issue is where this strategy will emerge from: IT or management?

Unfortunately, the role and morale of IT has been decimated over the past few decades. Their deep understanding of technology – *if* properly combined with a broad understanding of business issues – could be immensely beneficial to their companies.

The transition from *building* complex architectures to *buying*, and implementing pre-fab enterprise software, moved IT toward the periphery of business. And just as IT implementations were getting more costly and less productive, technology from home life was starting to permeate work life, giving workers access to intuitive and familiar user interfaces. Now IT needs to be able to shift from *buy* to *compose*, which means that IT needs to get cleverer with the connective capacities of various technologies.

The 'Cloud' is commoditizing information technology, which makes this moment make-or-break for the 'Department Previously known as IT'. Either they take the lead in crafting a technology strategy that drives technology-enabled innovation, or they go extinct.

WE'RE HALFWAY THERE

This book started with a simple observation on our journey into digital: we're probably only halfway there. Over the past year, I've delivered numerous lectures around the world on this subject, and I've come to the strange conclusion that two types of audiences often have difficulty with my statement: the Germans and the Accountants. These two demographics keep asking me over and over again: "But how do you know that we're *exactly* halfway?" I don't. This is basically an exercise to try and think what will happen once technology stops being technology, and digital becomes 'normal'.

My favorite thinker and author, certainly on technology, was the late Douglas Adams. A writer by profession, Douglas Adams was also a true technologist, and one of digital's biggest fans.

Although he passed away too early in life in 2001, he articulated with brilliant insight what would happen once 'technology stops being technology'.

> As a matter of fact, Douglas Adams had a first class definition of technology: "Technology is a word that describes something that doesn't work yet."

When questioned at a conference, in response to a panel of representatives of the music, publishing and broadcasting industries, who had asked him how he thought the technological changes would affect them, Douglas Adams said: "This is like a bunch of rivers, the Amazon and the Mississippi and the Congo asking me how the Atlantic Ocean might affect them... and the answer of course is that they won't be rivers anymore, just currents in the ocean."

This book is about the trend-spotting and creative thinking we need to try and fathom what will happen when we do reach that big Ocean of The New Normal, and when the old notion of 'rivers' suddenly seem irrelevant.

SO WHAT HAPPENS NEXT?

The second question I get asked a lot is: "What happens next? What happens beyond the New Normal?" Indeed, although we've explored some 'limits' that will affect us in the New Normal, it would be foolish to think that it will end there. So what lies even further ahead?

This is where the great minds of the TechnoFuturists fail to agree.

One of the most prominent futurists is Ray Kurzweil. Kurzweil believes that the advances of computer technology will follow an exponential pattern as opposed to a a linear one. This is an incredibly powerful distinction. If you were offered a job that paid $1 as a starting salary on January 1st, and told that your salary would increase by $1 (linearly) each successive day until the end of the month, I would think that you'd turn down the job. On the other hand, if you were offered a job that paid $1 as a starting salary, but that salary doubled each successive day until the end of the month, you would be earning more than a billion $ on January 31st. Not bad for a day's work.

Apply this powerful idea of exponential growth to technology and you see how quickly we are headed toward a fundamental paradigm shift even more disruptive than the New Normal. Kurzweil predicts that by 2029, a computer will demonstrate computing capacity indistinguishable from a human mind in terms of knowledge, emotion and self-awareness, thus passing the Turing test.

> Turing test:
> Alan Turing's 1950 description of a dialogue in which a person tries to guess which of two conversations is being conducted with a person and which with a computer.

By 2050, it will be possible to 'download' a complete human brain – with all of its accumulated memories and knowledge – onto a computer infrastructure.

The moment when computer technology could surpass human capabilities and then continue to accelerate the pace of intelligence creation, is called 'Singularity'.

Ray Kurzweil fundamentally believes that the Singularity is inevitable, and believes that it will happen around the year 2050.

Whether it will be a proper 'Singularity' or some other technological shift, there is very likely another S-curve right behind the one we've investigated in this book.

That means that there will very likely be another evolution where new and exciting 'technology' will seem special and weird at first, leading to a hype and a crash. And then again an era of 'normality' will settle in when this new technology stops being 'technology'. Kurzweil believes it is likely that the entire cycle after this happens faster.

I hope the ideas in this book have sparked your thinking about what will happen to your organization once we hit the New Normal, and how this will affect your employees and your customers. And I hope it has at least communicated the point that it is vital for all businesses – bits or atoms – to reflect on how their markets will shift once digital becomes normality.

The best (and easiest) way to think through these issues may be through the eyes of our children. If you have the chance to see your children grow up as digital natives, see how they breathe digital without even knowing that it is digital, see how they take digital for granted, then you know that a big change is coming. That gut feeling is something that needs to be internalized by your company as well. I've spent too much time in brainstorm meetings with senior executives who 'want to think about the future', while it would have been much better for them to listen to the new generation of digital natives, new recruits and young customers, and try to understand how this next generation of citizens has chosen to live and function.

I remember the article by Frances Cairncross, working for the Economist at the time, about 'The Death of Distance'. This article appeared in the early days of the first internet hype. Today we are fully aware of the Death of Distance, and communicating with people around the globe has become not only common practice, but has had a significant impact on the way we work and live. But the New Normal has another fatal casualty, the 'Death of Time'.

Recently I was browsing through the proceedings of an O'Reilly conference in Seattle when one of the names of the speakers caught my eye. "Annalee Newitz" rung a bell in the back of my head, because I had spent three years in class sitting next to a funky girl with exactly that name almost 25 years ago.

Googling her name took me a minute, and I found out that indeed this was the exact person I had been friends with so long ago. She had become an influential tech writer, writing for Wired magazine, covering the technology scene in Silicon Valley, and was now living in San Francisco. On her personal blog, of course, I not only found her complete history, but also links to appearances and speeches on YouTube, as well as a link to a one hour interview with her on the net. I downloaded the file on my phone and listened to it on the way home that night.

It was the most amazing experience. Here I was, driving home, listening to a voice I had not heard for 25 years, and a swarm of thoughts and experiences came flooding back from deep within my brain as the voice began to rake up my own history. It was like a time capsule of my youth triggered by the recording, and an immense feeling of the collapse of time. The next day I gmailed Annalee, and – thanks to the Death of Distance – the link was reestablished, closing a quarter-of-a-century in the blink of an eye.

Stranger still was the realization that for my kids, no such gaps will exist, because time capsules abound in a world where all information is captured.

How can we prepare for the New Normal? How can we best brace ourselves when the full tsunami power of the New Normal hits us?

In the New Normal, we will see the blending of boundaries between technology and business. We will see the creation of hybrid organizations where technology and business fuse together. But these hybrid opportunities will only be realized by people who can be multi-faceted, think broadly and deeply, and have the capacity to become T-shaped individuals.

In the 5th century B.C, the Greek philosopher Heraclites was famous for his doctrine of 'change' being central to the Universe: "All is flux, nothing stays still".

Change is a constant, and we will have to adapt.

I wish you a lot of luck in the New Normal.

SOURCES

CHAPTER 1

→ Peter Hinssen, *Business / IT Fusion. How to move beyond Alignment and transform IT in your organization*, Mach Media NV, 2008

→ John Naisbitt, *Megatrends. Ten New Directions Transforming Our Lives*, Warner Books, 1982

→ Nicholas Negroponte, *Being Digital*, Knopf, 1995

CHAPTER 2

→ Chris Anderson, *The long tail: how endless choice is creating unlimited demand*, Random House Business Books, 2006

→ Chris Anderson, *Free: The Future of a Radical Price*, Hyperion, 2009

→ Stewart Brand, *The Media Lab: Inventing the Future at MIT*, Viking Publishing, 1987

→ Barry Schwartz, *The Paradox of Choice: Why More Is Less*, Ecco, 2004

CHAPTER 3

→ Douglas Adams, *Salmon of Doubt: And Other Writings*, Macmillan, 2002

CHAPTER 4

→ Charles Mackay, *Extraordinary Popular Delusions & the Madness of Crowds*, with a foreword by Andrew Tobias, Harmony Books, 1980

→ Carlota Perez, *Technological Revolutions and Financial Capital: The Dynamics of Bubbles and Golden Ages*, Edward Elgar, 2002

→ Seth Godin, *Purple Cow: Transform Your Business by Being Remarkable*, Portfolio Hardcover, 2003

→ Marshall McLuhan, *Understanding Media: The Extensions of Man*, McGraw Hill, 1964

→ B. Joseph Pine II and James H. Gilmore, *The Experience Economy: Work Is Theater & Every Business a Stage*, Harvard Business School Press, 1999

→ Chris Anderson, *The Long Tail: Why the Future of Business is Selling Less of More*, Hyperion, 2006

→ Don Peppers and Martha Rogers, *The One to One Future*, Currency, 1996

CHAPTER 5

→ John Naisbitt, *Megatrends. Ten New Directions Transforming Our Lives*, Warner Books, 1982

→ Malcolm Gladwell, *What The Dog Saw: And other adventures*, Little, Brown and Company, 2009

→ E. O. Wilson, *Consilience: The Unity of Knowledge*, Knopf, 1998

→ Isaac Newton, *Philosophiæ Naturalis Principia Mathematica*, Latin for "Mathematical Principles of Natural Philosophy", often called the "Principia Mathematica", 1687

→ T.S. Eliot, *The Rock*, Faber & Faber, 1934

CHAPTER 6

→ Gifford and Elizabeth Pinchot, *Intrapreneuring: Why You Don't Have to Leave the Corporation to Become an Entrepreneur*, Harper & Row, 1985

→ Robert A. Burgelman, *Corporate Entrepreneurship and Strategic Management: Insights from a Process Study*, Management Science Vol. 29 n° 12, Dec. 1983

→ Timothy Ferriss, *The 4-Hour Workweek: Escape 9-5, Live Anywhere, and Join the New Rich*, Crown, 2007

→ Morten T. Hansen, *T-Shaped Stars: The Backbone of IDEO's Collaborative Culture, An Interview with IDEO CEO Tim Brown*, ChiefExecutive.net

CHAPTER 7

→ Vandana Singh and Amy Sonpal, *The Downfall of Polaroid: Corporate Lessons (Part A)*, IBS Case Development Centre, www.ibscdc.org, 2007

→ Mic Wright, *Kodak develops: A film giant's self-reinvention*, Wired UK, www.wired.co.uk, March 2009

→ Henry Chesbrough, *Open Innovation: The New Imperative for Creating and Profiting from Technology*, Harvard Business Press, 2003

→ Larry Huston and Nabil Sakkab, *P&G's New Innovation Model*, Harvard Business Review, hbswk.hbs.edu, March 2006

→ Leander Kahney, *How Apple Got Everything Right By Doing Everything Wrong*, Wired magazine, March 2008

CHAPTER 8

→ Nicholas Carr, *IT Doesn't Matter*, Harvard Business Review, hbr.org, May 2003

→ Nicholas Carr, *The Big Switch: Rewiring the World, from Edison to Google*, W. W. Norton & Company, 2008

→ *Smarter IT investments*, The McKinsey Quarterly Chart Focus Newsletter, www.mckinseyquarterly.com, October 2007

→ Michael Porter, *On Competition, Updated and Expanded Edition*, Harvard Business School Press, 2008

About Peter Hinssen's previous book,

BUSINESS/IT FUSION (2008)

Beautifully written, thoroughly researched, spiced with stories and a unique combination of humor and intellect. Peter challenges his readers to do everything they can to fuse IT and the rest of the business.

Susan Cramm
Founder and president of Valuedance, IT leadership coach, former CIO and CFO,
and award-winning author of the Harvard Business Press book, "8 Things We Hate about I.T."

Read this book if you are interested in repositioning "the department previously known as IT".

Rob Goffee
Professor of Organisational Behaviour, London Business School

Business/IT Fusion hits the nail right on the head. It is a great mental journey on how to move beyond mere alignment thinking, and really transform IT into a true business asset.

Steve Van Wyk
Head of Operations and IT Banking, ING Group

Business/IT Fusion belongs on the bookshelf of every CIO as well as every leader who aspires to transform the IT department.

Costas Markides
Professor of Strategy and holder of the Robert P. Bauman Chair in Strategic Leadership,
London Business School

PETER HINSSEN

An entrepreneur, advisor, lecturer and writer, Peter Hinssen (1969) is one of Europe's most sought-after thought leaders on the impact of technology on society and business. He is frequently called upon to lead seminars and consult on issues related to the adoption of technology by consumers, the impact of the networked digital society, and the Fusion between business and IT.

Peter has extensively researched the organizational transformation of IT departments, the profile of the next-generation CIOs and the profound shift in IT roles during a Fusion process.

Peter Hinssen is co-founder of Across Group and Chairman of Across Technology. He has been an Entrepreneur in Residence with McKinsey & Company, and Chairman of Porthus, one of Europe's first Software-as-a-Service companies.

Peter coaches business executives on developing future innovation perspectives, and is a board advisor on subjects related to innovation and IT.

He lectures on IT Strategy at various business schools in Europe such as London Business School (UK), TiasNimbas Business School (Netherlands) and Vlerick Leuven Gent Management School (Belgium). Peter is a passionate keynote speaker frequently welcomed at CIO forums and conferences around the world.

Business/IT Fusion, written and published in 2008, became an instant reference for IT and business executives around the globe. In his second book, **The New Normal** (2010), Peter demonstrates how companies should explore the limits of the digital world.

www.peterhinssen.com – www.neonormal.com

The concepts discussed in *The New Normal* and *Business/IT Fusion* are put into practice at **Across Technology** – Peter Hinssen's company. Across Technology focuses on areas such as IT strategy, transformation, governance, communication, knowledge management and CxO Coaching.

For more info: www.a-cross.com